Numerical Methods

NIPA® GENX ELECTRONIC RESOURCES & SOLUTIONS P. LTD.
New Delhi-110 034

About the Authors

Atul Goyal is a Head of Department at Petroleum Engineering Department at Guru Gobind Singh College of Engineering and Technology, Guru Kashi University, Talwandi Sabo (Bathinda). His qualification is B.Tech, M.Tech, MCA. He received a Dale Carnegie Training certificate in High Impact Teaching Skills from Wipro Mission10X and an appreciation letter from the Vice Chancellor of Guru Kashi University for his excellent work. He became a Rashtrapati Scout to serve the nation and his fellow humans. He has published several research papers in a number of journals, and books (Disaster Management and Engineering Chemistry). He has participated in number of workshops, and is a life member of the ISTE, member of the Board of Studies of Petroleum Engineering and editorial member of the Wyano Academic Journal & i-Xplore International Research Journal Consortium. He is also coordinator of the sports, scholarship & NSS and an active member of discipline committee.

Dr. Madhuchanda Rakshit, an Assistant Professor, presently working at Guru Kashi University, Talwandi Sabo (Bathinda). Completed B.Sc, B.Ed, M.Sc and Ph.D degrees from Gauhati University, Assam with excellence. Worked under an NBHM funded Project in IASST (Institute of Advanced study in Science and Technology), Guwahati. Earlier worked in Handique Girls College in Guwahati, Assam, Haldia Institute of Technology, Bharatiya Vidya Bhawan and Sikkim Manipal University in Haldia, West Bengal. More than thirty (30) research papers published on her own credit in various SCI and UGC referred journals. Editor of International Indological Education and Research Institute, Chandigarh, member of editorial board of Guru Kashi Journal of Research in Engineering and Technology, member of Board of Studies of Guru Kashi University and also associated with editorial board of some other journals. One Ph.D is awarded under her supervision and three (3) more scholars are doing Ph.D under her supervision and four (4) M.Phil degrees were awarded under her supervision. Her area of research constitutes Operation Research, Queueing Theory, Time series Analysis, Fractional and Quadratic Linear Programming problem, Transportation Problem etc. Her area of proficiency is not only in Applied Mathematics but also in Education and Management studies.

Dr. Suchet Kumar received Ph.D in Mathematics at Guru Kashi University. He has completed his Masters in Mathematics from Chaudhary Devi Lal University, Sirsa (Haryana) in 2011 by securing 78% marks. He has passed his graduation in B.Sc. (Electronics) from Govt. National College, Sirsa affiliated with Kurukshetra University, Kurukshetra by securing 72 % marks. He has qualified CSIR/ UGC NET (June-2012) with AIR-18, GATE-2012 with AIR-52, Rajasthan State Eligibility Test-2012 for Lectureship, Haryana Teacher Eligibility Test-2016 for Post Graduate Teacher. Previously, he has worked as Assistant Professor in S. Balraj Singh Bhunder Memorial University College, Sardulgarh (Distt. Mansa) for one year. Presently, he is working as teacher in Mathematics at Govt. Sen. Sec. School, Fatta Maloka (Distt. Mansa (Punjab)).

www.nipaers.com

Browse, Search, Read & Buy...

Print Books eChapters
eBooks Publishers
Forthcoming Subject Catalogues
New Titles

Online Resources on:

- ✓ Current Affairs
- ✓ Reasoning, Logic & Aptitude
- ✓ Competitive Examinations
- ✓ English Language Lab
- ✓ Writing & Pronunciation Tools
- ✓ Personality Development
- ✓ Online Programmes for Professionals
- ✓ Interview Preparation

Pay Using

CC Avenue®

Numerical Methods

Atul Goyal
Head of Department
Department of Petroleum Engineering
Guru Kashi University
Talwandi Sabo, Bathinda, Punjab

Madhuchanda Rakshit
Assistant Professor Mathematics
Guru Kashi University
Talwandi Sabo, Bathinda, Punjab

Suchet Kumar
Mathematics Teacher
Govt. Sen. Sec. School
Fatta Maloka, Distt., Mansa, Punjab

NIPA® GENX ELECTRONIC RESOURCES & SOLUTIONS P. LTD.
New Delhi-110 034

NIPA® GENX ELECTRONIC RESOURCES & SOLUTIONS P. LTD.

101,103, Vikas Surya Plaza, CU Block
L.S.C. Market, Pitam Pura, New Delhi-110 034
Ph : +91 11 27341616, 27341717, 27341718
E-mail: newindiapublishingagency@gmail.com
Web: www.nipabooks.com

For customer assistance, please contact
Phone: + 91-11-27 34 17 17
Fax: + 91-11- 27 34 16 16
E-Mail: feedbacks@nipabooks.com

© 2023, Publisher

ISBN: 978-81-19235-24-7

All rights reserved. No part of this publication may be reproduced, stored in a retrieval system or transmitted in any form or by any means, including electronic, mechanical, photocopying recording or otherwise without the prior written permission of the publisher or the copyright holder.

This book contains information obtained from authentic and highly reliable sources. Reasonable efforts have been made to publish reliable data and information, but the author/s, editor/s and publisher cannot assume responsibility for the validity, accuracy or completeness of all materials or information published herein or the consequences of their use. The work is published with the understanding that the publisher and author/s are not attempting to render any professional services. The author/s, editor/s and publisher have attempted to trace and acknowledge the copyright holders of all material reproduced in this publication and apologize to copyright holders if permission and/or acknowledgements to publish in this form have not been taken. If any copyrighted material has not been acknowledged, please write to us and let us know so that we may rectify the error, in subsequent reprints.

Trademark Notice: NIPA®, the NIPA® logos and their presentations (the way they are written/presented) in this book are the trademarks of the publisher and hence may not be used without written permission, if copied or used without authorization, the infringer will be prosecuted as per law.

NIPA® also publishes books in a variety of electronic formats. Some content that appears in print may not be available in electronic books, and vice versa.

Composed and Designed by NIPA®.

Dr. Neeraj Bansal
Department of Mathematics
Government Rajindra College
Bathinda, Punjab

Foreword

The title of this book reflects its structure and style with great exactness. Balance between theory and examples are maintained. The material of the book is in simple language. It can be used both as an introduction and a refinement to course. Not only the students in engineering, science, and mathematics course but also other advanced readers could experience this book as valuable one in clearing the basic fundamental concepts of numerical methods. I strongly recommend this book for the readers.

Neeraj Bansal

Preface

The objective of Numerical Methods is to solve complex numerical problems using simple arithmetic operations in order to develop and evaluate methods for computing numerical results from given data. Before the advent of modern computers numerical methods often depended on hand interpolation in large printed tables. In 20th century, after the advent of computer; numerical methods are the methods for solving problems on computers through calculations which results in table of numbers and/or graphical representations/ figures.

This book is written in simple and easy language, in systematic manner, student-friendly and numerical problem solving orientation. Emphasis should be given on examples rather than pure theory. The theory part should be kept in minimum and presented in a heuristic and intuituve manner. Each concept can be justified with the help of examples (which is unavailable in other books) as student may come dilemma to find the solution of the concept from other books. So learning is with the help of examples as examples are the best source to learn and remember that particular problem. At the end of chapters, excercise questions will be given.

This book is meant for undergraduate and graduate students of engineering, science and mathematics i.e., B.Tech (Chemical, Electrical, Production, Mechanical, Civil, Petroleum Engineering), BA (Math), B.Sc (Math), B.Sc (Computer Applications), BCA. The book is also useful for the instructors. In addition to this, the book is useful for engineers, managers and researchers to learn the basic principles and concepts of Numerical Methods for the purpose of design and analysis. The book can be adapted for short course on numerical methods.

Now, this book is in hands of the users. The authors are highly optimistic of receiving encouraging response from the readers. Any comments and suggestions from the readers for the further improvement of the book are highly appreciated and gratefully accepted.

Authors

Acknowledgement

This book would not have been possible without the love, guidance and support of our entire families. We would like to extend highly thanks to all our colleagues who helped us to prepare this dream into a concrete reality.

We thank the entire team of NIPA for their support and continuous encouragements and interest to publish this work.

All these thanks are only a small fraction of what is due to Almighty God, for giving us health, strength and wisdom to accomplish this assignment.

Er. Atul Goyal
Dr. Madhuchanda Rakshit
Dr. Suchet Kumar

Chapterwise Summary

Chapter 1 is Numerical Computation. Basic introduction along with the significance of numerical methods is mentioned in the chapter. In addition , in this chapter, Taylor's theorem, a few basic ideas and concepts regarding numerical computations, number representation, error consideration, various types of errors, general error formula including approximation of function, stability and condition, uncertainty in data and noise are described.

Chapter 2 is Linear System of Equations. It deals with the linear system of equations along with its solution. This chapter covers inverse of a matrix, matrix inversion method including augmented matrix, Gauss elimination method, Gauss-Jordan method, Cholesky's Triangularisation method, Crout's method, Thomas Algorithm for tridiagonal system, Jacobi's and Gauss-Seidal iteration method, Rayleigh's power method.

Chapter 3 is Solution of Algebraic and Transcendental Equations. It covers solution of non-linear equations. In this chapter, the various covered topics are bisection method, regula-falsi method, method of false position, simple and modified Newton-Raphson method, successive approximation method, second method, Muller's method, Chebyshev method, Aitken's Ä^2 method and comparison of various iterative methods.

Chapter 4 is Numerical Differentiation. The topics covered include the derivatives based on Newton's forward and backward interpolation formula, and based on Stirling's interpolation formula.

Chapter 5 is Finite Differences and Interpolation. It includes topics on finite, forward, backward and central differences. It also includes error propagation in difference table, properties of operator Ä, difference operators, relationship among the operators, representation of a polynomial using factorial notation, interpolation with equal and unequal intervals, central difference and Newton's divided difference interpolation formula and cubic spline interpolation.

Chapter 6 is Curve Fitting, Regression and Correlation. Topics discussed in this chapter are on linear equation, curve fitting with a linear equation, criteria for "best" fit, linear least-squares regression and method of least squares fitting (for straight line, 2nd degree parabola and exponential curves).

Chapter 7 is Numerical Integration. In this chapter presentation is given on topics on Newton-Cotes closed and Gauss-Legendre quadrature formula, Gaussian integration, Euler-Meclaurin formula, Trapezoidal rule, Simpson's 1/3 and 3/8 rule along with estimation of error, Boole's and Weddle's rules.

Chapter 8 is Numerical Solution of Ordinary and Partial Differential Equations. In this chapter, the various stressed topics are Picard's method of successive approximation, Taylor's series method, simple and modified Euler's method, Runge-Kutta method, Adams-Bashforth method, prediction-correction method, solution of laplace equation through Jacobi's and Gauss-Siedel method.

Contents

Foreword .... *vii*

Preface .... *ix*

Acknowledgement .... *xi*

Chapterwise Summary .... *xiii*

1. Numerical Computation 1

1.1 Introduction to Numerical Method 1

1.2 Significance of Numerical Method 2

1.3 Taylor's Theorem 4

1.4 Number System 5

1.4.1 Decimal Number System 6

1.4.2 Binary Number System 6

1.4.3 Octal Number System 7

1.4.5 Conversion from Base 10 to a New Base 7

1.4.6 Conversion from a Base Other Than 10 to a Base Other Than 10 8

1.4.7 Shortcut Method for Conversion From One Number System To Other 10

1.5 Recursive Algorithms for Conversion 11

1.6 Significant Digits 11

1.7 Error Consideration 13

1.7.1 Absolute Errors 13

1.7.2 Relative Errors 14

1.7.3 Inherent Errors 14

1.7.4 Round-off Errors 14

1.7.5 Truncation Errors 15

1.7.6 Systematic Errors 16

1.7.7 Random Errors 16

1.7.8 Error Propagation 16

1.8 Error Estimation 18
1.9 General Error Formula 19
1.9.1 Function Approximation 21
1.9.2 Stability and Condition 22
1.9.3 Uncertainty in Data or Noise 23

2. Linear System of Equations 25

2.1 Introduction
2.1.1 Introduction to Determinants 25
2.1.2 Introduction to Matrices 27
2.2 Methods of Solution 32
2.3 The Inverse of a Matrix 33
2.4 Cramer's Rule 35
2.5 Matrix Inversion Method 38
2.5.1 Augmented Matrix 42
2.6 Gauss Elimination Method 44
2.7 Gauss-Jordan Method 52
2.8 Crout's Method 55
2.9 Cholesky's Triangularisation Method 62
2.10 Thomas Algorithm for Tridiagonal System 66
2.11 Jacobi's Iteration Method 71
2.12 Gauss-Seidal Iteration Method 77
2.13 Iterative Method for Eigen Values and Eigen Vectors 82
2.13.1 Jacobi's Method 83
2.13.2 Reyleigh's Power Method 87

3. Solution of Algebric and Transcendental Equations 93

3.1 Introduction 93
3.2 Solution of Algebraic and Transcendental Equations 94
3.3 Bisection (or Bolzano) Method 95
3.4 Regula Falsi or False Position Method 98
3.5 Fixed Point Iteration Method 102
3.6 Iteration Method 103
3.7 Newton-Raphson Method 104
3.7.1 Convergence of Newton-Raphson Method 107
3.7.2 Rate of Convergence of Newton Raphson Method 110

3.7.3 Modified Newton-Raphson Method 111
3.7.4 Rate of Convergence of Modified Newton-Raphson Method 113
3.8 Successive Approximation Method 114
3.8.1 Error Estimate in the Successive Approximation Method ... 115
3.9 Secant Method 116
3.9.1 Convergence of the Secant Method 118
3.10 Muller's Method 119
3.11 Chebyshev Method 122
3.12 Aitken's D2 Method 123
3.13 Comparison of Iterative Methods 125
3.14 Solution of Non-Linear Equations 126
3.14.1 Iteration Method 126
3.14.2 Newton-Raphson Method 128

4. Numerical Differentiation 133
4.1 Introduction 133
4.2 Derivatives Based on Newton's Forward Interpolation Formula . 133
4.3 Derivatives Based on Newton's Backward Interpolation Formula 135
4.4 Derivatives Based on Stirling's Interpolation Formula 138
4.5 Derivatives Based on Bessel's Interpolation Formula 141
4.6 Derivatives Based on Newton's Divided Difference Formula 143

5. Finite Differences and Interpolation 147
5.1 Introduction 147
5.2 Finite Difference Operators 147
5.2.1 Forward Differences 147
5.2.2 Backward Differences 149
5.2.3 Central Differences 151
5.2.4 Error Propagation in Difference Table 152
5.2.5 Other Difference Operator 154
5.2.6 Relationship between the Operators 154
5.2.7 Representation of a Polynomial using Factorial Notation 156
5.3 Interpolation with Equal Intervals 161
5.3.1 Newton's Forward Interpolation Formula 161
5.3.2 Newton's Backward Interpolation Formula 165

5.4 Interpolation with Unequal Intervals 167
5.4.1 Lagrange's Interpolation Formula 167
5.4.2 Newton's General Interpolation Formula with Divided Differences 169
5.5 Central Difference Interpolation Formula 172
5.5.1 Gauss Forward Interpolation Formula 172
5.5.2 Gauss Backward Interpolation Formula 174
5.5.3 Stirling's Formula 176
5.5.4 Bessel's Formula 176
5.5.5 Laplace-Everett's Formula 178
5.6 Cubic Spline Interpolation 178

6. Curve Fitting 187
6.1 Introduction 187
6.2 Linear Equation 188
6.3 Curve Fitting with Linear Equation 191
6.4 Criteria for a Best Fit 192
6.5 Linear Least-Square Regression 193
6.6 Method of Least Squares Fitting: Straight Line, 2nd Degree Parabola, Exponential Curves 194
6.6.1 Fitting of straight line 194
6.6.2 Fitting of second degree parabola 195
6.6.3 Fitting of exponential curve 197

7. Numerical Integration 201
7.1 Introduction 201
7.1.1 Relative Error 202
7.2 Newton-Cotes Closed Quadrature Formula 203
7.3 Gauss-Legendre Quadrature Formulae 205
7.4 Gaussian Integration 211
7.5 Euler-Meclaurin Formula 212
7.6 Trapezoidal Rule 214
7.6.1 Error Estimate in Trapezoidal Rule 217
7.7 Simpson's 1/3 Rule 218
7.7.1 Error Estimate in Simpson's 1/3 Rule 222
7.8 Simpson's 3/8 Rule 225

7.9 Boole's and Weddle's Rules 230
7.9.1 Boole's Rule 230
7.9.2 Weddle's Rule 236

8 Numerical Solution of Ordinary and Partial Differential Equations 251

8.1 Introduction 251
8.2 Picard's Method of Successive Approximation 253
8.3 Taylor's Series Method 255
8.4 Euler's Method 257
8.5 Modified Euler's Method 259
8.6 Runge-Kutta Method 261
8.7 Adams-Bashforth Method 270
8.8 Prediction-Correction Method 273
8.9 Boundary Value Problems 276
8.10 Solution of Laplace Equation - Jacobi's Method, Gauss-Siedel Method 276

Index 289

1

Numerical Computation

1.1 Introduction to Numerical Method

There are three types by which problem / system can be solved and are analytical methods, graphical methods and numerical methods.

(a) **Analytical methods:** This approach or methods are used to find the exact solution of simple problems / systems rather than complex systems. It provides the behavior and characteristic features of a particular system, but can only be used for systems which can be approximated through use of linear models. Another limitation is in the case of practical problems, consisting of complex and nonlinear equations / processes, as analytical methods cannot handle properly.

(b) **Graphical methods:** This approach or methods are used to find the approximate solution of problems / systems through drawing of graphs for each of the equation and then check for the intersection of graphs. Intersection of graphs represents solution to system of equations, and no interaction or parallel lines of graphs represents no solution. This is not useful because of non-precise nature due to graphs drawing through hands, which also makes the process tedious. Another limitation is its viability for problems involving three or lesser dimensions.

(b) **Numerical methods:** This approach or methods are used to find the approximate solution of problems / systems in a simpler way involving a large quantity of long arithmetic calculations, as compared to analytical methods. Numerical methods are the techniques by which mathematical problems are formulated to find the solution through arithmetic operations.

Previously analytical and graphical methods were preferred because of time consuming nature of numerical methods, due to performing arithmetic operations through hand tools such as paper, pencil, calculator and slide rules, which ultimately results in wrong solution, called human error. But with the advent of affordable computers in this era shifted the paradigm from analytical and graphical methods to numerical methods as millions of calculation can be done in fractions of seconds. Numerical methods (NM) now becomes indispensible tool in the hands of scientists and engineers through high speed computers and increasing demand in order to provide answers / solutions through conservative methods of NM.

It also depends upon the input which is rarely exact because it comes from some measurement or method or from both, and further the method enhances the error resulting in wrong or inappropriate solutions. So in this chapter, concentration is on errors and its types, approximation, propagation and estimation.

1.2 Significance of Numerical Method

Numerical methods are the algorithms that provide approximate results of the problem. It is important to recognize the starting value's and must ensure the selected method shall work / converge and provide solution. It can be used for

(a) finding the roots of equation i.e., matrix inversion method, Gauss elimination method, Gauss Jorden method, Crout's method, bisection method, iteration method, method of false position, Newton-Raphson method, Secant method etc,

(b) solving ordinary and partial differential equations i.e., Picard's method, Euler's method, Runga-Kutta method, Adams-Bashforth method etc,

(c) finding integration values i.e., Gaussian integration, Trapezoidal rule, Simpson's rule, Bool's and Weddle's rule etc,

(d) finite differences and interpolation i.e., Newton interpolation, Cubic Spline interpolation etc,

(e) finding differentiation values i.e., Newton formula, Stirling's formula etc.

Real life examples

(1) **Traffic problem:** After working hours in office, everyone is hurry to reach his/her home by spending minimum amount of time in traffic jam on roads and each driver has its own behavior which depends upon the destination, time and speed avoiding accidents. Each driver interacts with other driver. So mathematical model is required which can provide the understanding necessary to make the driver's life more pleasant.

(2) **Probability application:** In this, an insect is placed in any one of intersection (from 1-10) in the maze. If insect comes out from the maze,

it will not return to the maze and within the intersections the path chosen will be random. So what is the probability of an insect to get the food?

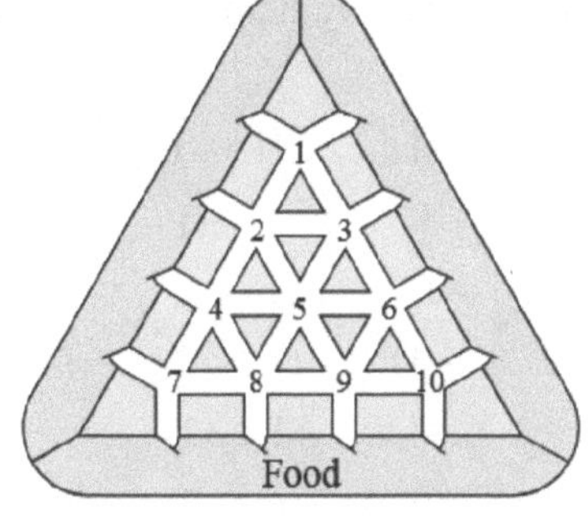

Fig. 1.1: Maze for an insect to find food

At intersection 1, the probability of finding food is ¼ of selecting upper path is lose i.e., ¼ (0) and on selection of lower path towards intersection 2, then probability is win i.e., ¼ P_2. Similarly towards intersection 3, probability is win i.e., ¼ P_3.

Overall at intersection 1, probability $P_1 = \frac{1}{4}(0) + \frac{1}{4}(0) + \frac{1}{4}(P_2) + \frac{1}{4}(P_3)$

Similarly at intersection 2, probability $P_2 = 1/5\,(0) + 1/5\,(P_1) + 1/5\,(P_3) + 1/5\,(P_4) + 1/5\,(P_5)$

$P_3 = 1/5\,(0) + 1/5\,(P_1) + 1/5\,(P_2) + 1/5\,(P_5) + 1/5\,(P_6)$

$P_4 = 1/5\,(0) + 1/5\,(P_2) + 1/5\,(P_5) + 1/5\,(P_7) + 1/5\,(P_8)$

$P_5 = 1/6\,(P_2) + 1/6\,(P_3) + 1/6\,(P_4) + 1/6\,(P_6) + 1/6\,(P_8) + 1/6\,(P_9)$

$P_6 = 1/5\,(0) + 1/5\,(P_3) + 1/5\,(P_5) + 1/5\,(P_9) + 1/5\,(P_{10})$

$P_7 = 1/4\,(0) + 1/4\,(1) + 1/4\,(P_4) + 1/4\,(P_8)$

$P_8 = 1/5\,(1) + 1/5\,(P_4) + 1/5\,(P_5) + 1/5\,(P_7) + 1/5\,(P_9)$

$P_9 = 1/5\,(1) + 1/5\,(P_5) + 1/5\,(P_6) + 1/5\,(P_8) + 1/5\,(P_{10})$

$P_{10} = 1/4\,(0) + 1/4\,(1) + 1/4\,(P_6) + 1/4\,(P_9)$

Rewriting the above 10 equations as

$$
\begin{array}{cccccccccccc}
4P_1 & -P_2 & -P_3 & & & & & & & & = & 0 \\
-P_1 & +5P_2 & -P_3 & -P_4 & -P_5 & & & & & & = & 0 \\
-P_1 & -P_2 & +5P_3 & & -P_5 & -P_6 & & & & & = & 0 \\
 & -P_2 & & +5P_4 & -P_5 & & -P_7 & -P_8 & & & = & 0 \\
 & -P_2 & -P_3 & -P_4 & +6P_5 & -P_6 & & -P_8 & -P_9 & & = & 0 \\
 & & -P_3 & & -P_5 & +5P_6 & & & -P_9 & -P_{10} & = & 0 \\
 & & & -P_4 & & & +4P_7 & -P_8 & & & = & 0 \\
 & & & -P_4 & -P_5 & & -P_7 & +5P_8 & -P_9 & & = & 0 \\
 & & & & -P_5 & -P_6 & & -P_8 & +5P_9 & -P_{10} & = & 0 \\
 & & & & & -P_6 & & & -P_9 & +4P_{10} & = & 0
\end{array}
$$

Hence, the augmented matrix is

$$\begin{bmatrix} 4 & -1 & -1 & 0 & 0 & 0 & 0 & 0 & 0 & 0 & 0 \\ -1 & 5 & -1 & -1 & -1 & 0 & 0 & 0 & 0 & 0 & 0 \\ -1 & -1 & 5 & 0 & -1 & -1 & 0 & 0 & 0 & 0 & 0 \\ 0 & -1 & 0 & 5 & -1 & 0 & -1 & -1 & 0 & 0 & 0 \\ 0 & -1 & -1 & -1 & 6 & -1 & 0 & -1 & -1 & 0 & 0 \\ 0 & 0 & -1 & 0 & -1 & 5 & 0 & 0 & -1 & -1 & 0 \\ 0 & 0 & 0 & -1 & 0 & 0 & 4 & -1 & 0 & 0 & 1 \\ 0 & 0 & 0 & -1 & -1 & 0 & -1 & 5 & -1 & 0 & 1 \\ 0 & 0 & 0 & 0 & -1 & -1 & 0 & -1 & 5 & -1 & 1 \\ 0 & 0 & 0 & 0 & 0 & -1 & 0 & 0 & -1 & 4 & 1 \end{bmatrix}$$

Through the help of Gauss-Seidel method by selection of $P_1 = P_2 = \ldots.. = P_{10} = 0$, the results after 19th iteration are; $P_1 = 0.090$, $P_2 = 0.180$, $P_3 = 0.180$, $P_4 = 0.298$, $P_5 = 0.333$, $P_6 = 0.298$, $P_7 = 0.455$, $P_8 = 0.522$, $P_9 = 0.522$, $P_{10} = 0.455$.

Other examples are root algorithm in computer science engineering, finding profit & loss of company, network simulation, weather prediction, climate models, train / traffic signal etc.

1.3 Taylor's Theorem

This theorem can be used to approximate the functions through polynomials and hence provides an estimate error in this operation. Taylor's theorem was devised by Taylor in 1712.

Suppose f be an (n+1) times differentiable function on an open interval containing points a and x, then

$$f(x) = f(a) + f'(a)\,(x-a) + \frac{f^2(a)}{2!}(x-a)^2 + \ldots. + \frac{f^n(a)}{n!} f^n(a)\,(x-a)^n + R_n\,(x)$$

where $R_n = \dfrac{f^{n+1}(c)}{(n+1)!}(x-a)^{n+1}$ for some number c between a and x.

This method is single step method and it works as successive derivatives, can be calculated. If function is complicated, then calculation of higher order derivatives become tedious and hence less importance, further it has little application in numerical computation. But this method provide initial values for the application of powerful methods such as Runga-Kutta method and Adams-Bashforth method.

Example 1.1: Solve y' = x + y, y(0) = 1 through Taylor's method. Find value of y at x = 0.1.

Solution: The given equation is:: y' = x + y

So

1st derivative:	$y' = x + y$	y (at x=0) or $y(0) = 1$
2nd derivative:	$y'' = 1 + y'$	$y''(0) = 2$
3rd derivative:	$y''' = y''$	$y'''(0) = 2$
4th derivative:	$y'''' = y'''$	$y''''(0) = 2$

etc.

So Taylor's series method is

$$y = y_0 + (y')_0 (x-x_0) + \frac{(y'')_0}{2!}(x-x_0)^2 + \frac{(y''')_0}{3!}(x-x_0)^3 + \ldots\ldots\ldots\ldots$$

Here $x_0 = 0$, $y_0 = 1$, and put the values

$$y = 1 + (1)(x) + \frac{2}{2!}x^2 + \frac{2}{3!}x^3 + \frac{2}{4!}x^4 + \ldots\ldots\ldots\ldots$$

When x = 0.1, then y is

$$y = 1 + (1)(0.1) + \frac{2}{2!}(0.1)^2 + \frac{2}{3!}(0.1)^3 + \frac{2}{4!}(0.1)^4 + \ldots\ldots\ldots\ldots$$

y = 1.1103

1.4 Number System

In information processing system, the number system provides the basis of all the operations. These number system are divided into groups of symbols such as 10 decimal digits, 26 alphabets of English, etc. In this positional number system, there are only a few symbols called digits, and these symbols represent different value depending on the position they occupy in the number. The value of each digit in such a number is determined by

(i) the digit itself,

(ii) the position of the digit in the number, and

(iii) the base of the number system, where base is defined as the total number of digits available in the number system.

There are two characteristics of all the number systems that are suggested by the value of the base, which determines the total number of different symbols or digits available in the number system. The first characteristics of these choices are always zero. The second characteristic is that the maximum value of a single digit is always equal to one less than the value of the base.

1.4.1 Decimal Number System

Decimal number system has played an important role in the development of science and technology. This number system is in use in our daily life, where base is equal to 10 because there are altogether 10 symbols or digits i.e., 0, 1, 2, 3, 4, 5, 6, 7, 8, 9 used in this system.

This decimal number system is based on the value of each digit which depends upon the position in a number. Each position is 10 times more significant than the previous one. The numeric value can be determined through multiplication of each digit of the number with the position value in which the digit appears and then adding the products. For example, 3456 is written as

$$2 * 1000 + 4 * 100 + 5 * 10 + 6 = 2000 + 400 + 50 + 6$$

1.4.2 Binary Number System

It is exactly same as that of decimal number system, except that the base is 2 instead of 10. There are only two symbols or digits i.e., 0 and 1.Each position in the system represents a power of the base 2. In the system, the rightmost position is the units (2^0) position, the second position from the right is the 2's (2^1) and so on. For example the decimal equivalent of the binary number 10101 is

$$(1*2^4) + (0*2^3) + (1*2^2) + (0*2^1) + (1*2^0) = 16 + 0 + 4 + 0 + 1 = 21$$

Therefore

$$(10101)_2 = (21)_{10}$$

Table 1.1: Binary number with their decimal value

Binary	Decimal equivalent
000	0
001	1
010	2
011	3

Contd.

100	4
101	5
110	6
111	7

Computer read binary number system in the form of 0's and 1's and is very long, so to provide short notation, octal and hexadecimal number systems are used.

1.4.3 Octal Number System

In this system, the base of radix of 8 is used i.e., digits are 0, 1, 2, 3, 4, 5, 6, 7. Similar to that of decimal number system, each position is 8 times more significant than the previous one i.e., each position in this system represents a power of the base 8. For example, the octal number 2057 is written as $(2057)_8$ and in the decimal form is

$(2*8^3) + (0*8^2) + (5*8^1) + (7*8^0) = 1024 + 0 + 40 + 7 = 1071$

Therefore

$(2057)_8 = (1071)_{10}$

1.4.4 Hexadecimal Number System

In this system, the base of radix of 16 is used i.e., digits are 0, 1, 2, 3, 4, 5, 6, 7, 8, 9, 10, A, B, C, D, E, F; where 0 to 9 are decimal number system and the remaining are number digits and represents 10, 11, 12, 13, 14, 15 respectively. The number digit A has a decimal equivalent of 10 and the hexadecimal F has a decimal equivalent value of 15. Thus the largest single digit is F or 15, which is one less than the base 16. For example, the decimal equivalent of the hexadecimal number IAF (written as IAF_{16}) is

$(IAF)_{16} = (1 * 16^2) + (A * 16^1) + (F * 16^0)$

$= (1 * 256) + (10 * 16) + (15 * 1)$

$= 256 + 160 + 15$

$= 431$

Therefore

$(IAF)_{16} = (431)_{10}$

1.4.5 Conversion from Base 10 to a New Base

The following four steps are used to convert a number from base 10 to a new base:

Step 1: Divide the decimal number to be converted by the value of the new base.

Step 2: Record the remainder from step 1 as the rightmost digit (least significant digit) of the new base number.

Step 3: Divide the quotient of the previous divide by the new base.

Step 4: Record the remainder from step 3 as the next digit (to the left) of he new base number.

Repeat steps 3 and 4, recording remainders from right to left, until the quotient becomes zero in the step 3. The last remainder obtained will be the most significant digit of the new base number.

Example 1.2: $25_{10} = x_2$

New base i.e., 2	Old base value	Remainders
2	42	
	21	0
	10	1
	5	0
	2	1
	1	0
	0	1

Hence $42_{10} = 101010_2$

Example 1.3: $952_{10} = x_{16}$

New base i.e., 16	Old base value	Remainders
16	428	
	26	12 = C
	1	10 = A
	0	1 =1

Hence $952_{10} = 1AC_{16}$

1.4.6 Conversion from a Base Other Than 10 to a Base Other Than 10

The following four steps are used to convert a number from other base 10 to a new other base than 10:

Step 1: Convert the original number to a decimal number of base 10.

Step 2: Convert the decimal number so obtained to the new base.

Example 1.4: $545_6 = x_4$

Step 1: Convert from base 6 to base 10

$545 = 5*6^2 + 4*6^1 + 5*6^0$

$= 5*36 + 4*6 + 5*1$

$= 209$

$545_6 = 209_{10}$

Step 2: Convert 209_{10} to base 4

New base i.e., 4	Old base value	Remainders
4	209	
	52	1
	13	0
	3	1
	0	3

Hence $209_{10} = 3101_4$

Therefore

$545_6 = 3101_4$

Example 1.5: $101110_2 = x_8$

Step 1: Convert from base 2 to base 10

$101110 = 1*2^5 + 0*2^4 + 1*2^3 + 1*2^2 + 1*2^1 + 0*2^0 = 46_{10}$

$101110_2 = 46_{10}$

Step 2: Convert 46_{10} to base 8

New base i.e., 8	Old base value	Remainders
8	46	
	5	6
	0	5

Hence $46_{10} = 56_8$

Therefore

$101110_2 = 56_8$

1.4.7 Shortcut Method for Conversion From One Number System To Other

(a) Shortcut method for binary to octal conversion

Step 1: Divide the binary digits into groups of three (starting from the right).

Step 2: Convert each group of three binary digits into one octal digit. Since decimal digits 0 to 7 are equal to octal digits 0 to 7 so binary to decimal conversion can be used in this step.

Example 1.6: $101110_2 = x_8$

Step 1: $\underline{101}\ \underline{110}$

Step 2: $101_2 = 1*2^2 + 0*2^1 + 1*2^0 = 5_8$

$110_2 = 1*2^2 + 1*2^1 + 0*2^0 = 6_8$

Hence, $101110_2 = 56_8$

(b) Shortcut method for octal to binary conversion

Step 1: Convert each octal digit to a three digit binary number

Step 2: Combine all the resulting binary groups (of three digits each) into a single binary number.

Example 1.7: $562_8 = x_2$

Step 1: $5_8 = 101_2$

$6_8 = 110_2$

$2_8 = 010_2$

Step 2: $568_2 = \underline{101}\ \underline{110\ 010}$

Hence, $562_8 = 101110010_2$

(c) Shortcut method for binary to hexadecimal conversion

Step 1: Divide the binary digits into groups of four (starting from the right).

Step 2: Convert each group of four binary digits into one hexadecimal digit.

Example 1.8: $11010011_2 = x_{16}$

Step 1: $\underline{1101}\ \underline{0011}$

Step 2: $1101_2 = 1*2^3 + 1*2^2 + 0*2^1 + 1*2^0 = 13_{10} = D_{16}$

$0011_2 = 0*2^3 + 0*2^2 + 1*2^1 + 1*2^0 = 3_{16}$

Hence, $11010011_2 = D3_{16}$

(d) Shortcut method for hexadecimal to binary conversion

Step 1: Convert the decimal equivalent of each hexadecimal digit to a four digit binary number.

Step 2: Combine all the resulting binary groups (of four digits each) into a single binary number.

Example 1.9: $2AB_{16} = x_2$

Step 1: $2_{16} = 2_{10} = 0010_2$

$A_{16} = 10_{10} = 1010_2$

$B_{16} = 11_{10} = 1011_2$

Step 2: $2AB_{16} = \underline{0010}\ \underline{1010}\ \underline{1011}$

Hence, $2AB_{16} = 001010101011_2$

1.5 Recursive Algorithms for Conversion

Recursive programming is so powerful because it maps so easily to proof by induction, making it easy to design algorithms and prove them correct. The drawback is that the recursion is not strongly supported by number of programming languages. So it is important to translate recursive algorithm into iterative algorithms, which yields lean and fast results. The steps are

(a) Function study

(b) Conversion of recursive calls into tail calls

(c) Introduction of loop around the function body

(d) Conversion of tail calls into continue statements

(e) Tidy up

1.6 Significant Digits

Significant figures in the measured value of any quantity tell the number of digits in which we have confidence. Larger the number of significant figures obtained, greater is the accuracy of the measurement. The reverse is also true.

For example, in case of wood thickness through vernier caliper may be $t_1 = 0.23$ cm (with 2 significant figures) and through spherometer it is $t_2 = 0.237$ cm (with 3 significant figures). Consider the last numeric is unreliable, the maximum error in t_1 is 0.23 ± 0.01 cm and the maximum error in t_2 is 0.237 ± 0.001 cm.

Percentage error in t_1 is $\left(\frac{0.01}{0.23} X100\right) = 4.34\%$ and percentage error in t_2

is $\left(\frac{0.001}{0.237} X100\right) = 0.422\%$. Hence 2nd observation through spherometer is more accurate than measurement from vernier caliper.

Rules for counting significant figures

(a) All non zero digits are significant and all zeros on the right of non zero digit are not significant. For example, x = 123 has 3 significant digits, y = 2100 has 2 significant digits.

(b) All zeros occurring between two non zero digits are significant. For example, x = 1004 has 4 significant digits.

(c) For number less than 1, all zeros between decimal point and the preceding non zero digit are not significant. For example, For example, x = 0.0012 has 2 significant digits, y = 1.001 has 4 significant digits as per point (b).

(d) All zeros on the right of the last non zero digit in the decimal part are significant. For example, x = 0.00800 has 3 significant digits, y = 1.00 has 3 significant digits.

(e) All zeros on the right of the last non zero digit become significant, when they come from a measurement. For example, x = 305 m as it has 3 significant digits. This can be expressed as 0.305 km or 0.305 X 10^5 cm with 3 significant digits.

Rules of rounding off

(a) If the digit is to be dropped is less than 5 then the preceding digit is left unchanged For example, x = 7.82 is rounded off to 7.8.

(b) If the digit is to be dropped is more than 5 then the preceding digit is increased by one. For example, x = 9.87 is rounded off to 9.9.

(c) If the digit is to be dropped is 5, which is followed by non zero, then the preceding digit is increased by one. For example, x = 19.852 is rounded off to 19.8.

(d) If the digit is to be dropped is 5 or followed by zero, then the preceding digit is left unchanged, if it is even. For example, x = 19.850 is rounded off to 19.8.

(e) If the digit is to be dropped is 5 or followed by zero, then the preceding digit is increased bu one, if it is odd. For example, x = 19.750 is rounded off to 19.8.

Accuracy: It is the closeness between the measured (or calculated) value and true value.

$$\text{Accuracy} = \frac{\text{True value - Measured value}}{\text{True value}}$$

Measurement of accuracy: It is an instrument's ability to measure true value up to within some stated error specifications.

Numerical accuracy: It is degree to which the numerical solution to the approximate physical problem approximates the exact solution to the approximate physical problem.

1.7 Error Consideration

Errors are usually regarded as negative, but mathematics error come in useful in statistics, computer programming, advanced mathematics and much more. Measuring process is a process of comparison. No measurement is perfect as the errors involved in the process cannot be completely removed. So **error** is the difference between measured (or calculated or approximate) value with the true (or exact) value. More is the difference, more is the error and less is the accuracy. So evaluating errors provides significantly useful information, especially when chance and probability is required.

Errors are of following types

(a) Absolute errors

(b) Relative errors

(c) Inherent errors

(d) Round-off errors

(e) Truncation errors

(f) Systematic errors

(g) Random errors

1.7.1 Absolute Errors

It is the error which is expressed in physical units. It is the difference between measured (or approximate) value and true value (or the average value) of a quantity. It is represented by E_A.

Absolute error (E_A) = True value – Measured value

1.7.2 Relative Errors

It is the error which is expressed as the fraction of absolute error to the true value (or the average value). It is expressed as E_R.

$$\text{Relative error } (E_R) = \frac{\text{True value - Measured value}}{\text{True value}} = \frac{E_A}{\text{True value}}$$

Relative error is always expressed in percentage error (E_P).

$$\text{Percentage error } (E_P) = 100 \text{ X } E_R = \frac{\text{True value - Measured value}}{\text{True value}} \text{X } 100$$

1.7.3 Inherent Errors

Errors which are already present in the statement of the problem, before the solution of it. These errors are due to limitation of mathematical tables, measuring or calculating device or due to approximate the given data. These are minimized through the use of high precision computing aids / devices.

1.7.4 Round-off Errors

During the computation / calculations, round-off errors appears due to the rounding off process. Rounding off originates due to limitation of computing aids i.e., a calculating device is capable of supplying only a certain number of digits in operations such as addition, subtraction, multiplication, division, conversion between number systems and so on. These errors can be minimized through changing in calculation procedure avoiding subtraction of nearly equal numbers, especially decimal point numbers, & division by small numbers and retain one more significant digit at each step than required significant digits & round off at the last step.

Types of round-off errors on a finite precision computing device are

(a) Negligible addition: When 2 numbers of different magnitudes are added or subtracted yielding largest number.

(b) Creeping round-off: Repeated rounding to n significant digits yielding error accumulation.

(c) Error magnification: It occurs when an enormous number is multiplied / divided with large / small magnitude.

(d) Subtractive cancellation: On subtraction of 2 nearly equal numbers where the difference exist in significant digits beyond the capacity of device to record it.

Suppose a = 237.6581, b = 238.2389, c = 0.014789, d = 137469, A = 238.0, B = 238.1, C = 0.01480, D = $1.375*10^5$.

On performing arithmatic operations manualy and which is rounded to 4 significant digits as follows

$a - c = 237.643203 \approx 237.6$

$b + d = 1377103.8389 \approx 1.377*10^5$

$b * d = 32697140.62 \approx 3.270*10^7$

$a / c = 15953.42015 \approx 1.595*10^4$

$a - b = -0.1808$

On performing arithmatic operations through device and which is rounded to 4 significant digits as follows

$A - C = 238.0$

$B + D = 1.377*10^5$

$B * D = 3.274*10^7$

$A / C = 1.608*10^4$

$A - B = -0.1$

Calculations through manually and device yields the following observations:

(a) Negligible addition: It has been observed that the round-off error has crept into the 4th significant digit on comparing operations a – c to A – C, and there is difference in 4th significant digit on comparing operations b + d and B + D.

(b) Creeping round-off: It has been observed that there is difference in 4th significant digit on comparing operations a / c and A / C, and two different answers on comparing b * d and B * D.

(c) Error magnification: It has been observed in operations a – c, b * d, a / c, b + d, the creeping round-offnad yielding gradual loss of precision as repeated rounding error accumulate.

(d) Subtractive cancellation: It has been observed that the introduction of significant error by working in fixed precision arithmetic on comparing operation a – b and A – B.

1.7.5 Truncation Errors

These are type of algorithm error and arise due to replacement of infinite series into finite one and usage of approximate results.

For example, if $x = 15.387$, $x' = 15.38$ (data required upto 4 digit), then

Truncation error = $x - x' = 15.387 - 15.38 = 0.007$

Difference between truncation and rounding off

For example, $x = 15.387$, for further usage if required 4 significant digits then truncation gives 15.38, but rounding off gives 15.39.

1.7.6 Systematic Errors

These are those errors whose causes are known, so these can be minimized. For example

(a) Machine errors / instrumental errors: It spoils calculations and procedure due to negative factors. Such as in case of electronics, due to failure of capacitor results in presence of DC voltage component and hence produces error. This can be removed through calibration and controls upon understanding or recognizing the problem.

(b) Personal errors: These arise due to observer's inexperience, such as lack of proper setting of apparatus, taking reading without proper precautions etc.

(c) Imperfection due to ignorance of certain facts such as effect of radiation, ignored generally, on heating the substance etc.

(d) External factors such as temperature, pressure, heat, humidity etc during the experiment, so need to apply some corrective action in order to minimize the effect.

1.7.7 Random Errors

This error is always present in the system and their causes are not known precisely. This error varies unpredictably in size as well as in direction. It is not possible to eliminate the random errors.

Random errors are sometimes called '**chance errors**'. For example, when the same person repeats the same observation, he may get different readings every time. So these are minimized by performing large number of observations and take the arithmetic mean of all the observations.

1.7.8 Error Propagation

The result of an experiment is calculated through performing mathematical operations, such as addition, subtraction, multiplication, division etc, on number of measurements results in different degree of accuracy. So to calculate the net error in the result, there is need to study how errors propagate in different mathematical operations.

Suppose X and Y are two true values and their approximate (or measured) values are X' and Y' respectively.

Absolute error for X i.e., $E_{AX} = X - X'$

$$\text{or } X = X' + E_{AX}$$

Absolute error for Y i.e., $E_{AY} = Y - Y'$

$$\text{or } Y = Y' + E_{AY}$$

(a) Error in addition operation

$X + Y = (X' + E_{AX}) + (Y' + E_{AY}) = X' + Y' + E_{AX} + E_{AY}$

$|(X + Y) - (X' + Y')| = | E_{AX} + E_{AY}| \leq | E_{AX}| + |E_{AY}|$

So the absolute error in taking (X' + Y') as an approximation to (X + Y) is less than or equal to the sum of the absolute errors in taking X' as approximation to X and Y' is an approximation to Y.

(b) Error in subtraction operation

$X - Y = (X' + E_{AX}) - (Y' + E_{AY}) = X' - Y' + E_{AX} - E_{AY}$

$|(X - Y) - (X' - Y')| = | E_{AX} - E_{AY}| \leq | E_{AX}| + |E_{AY}|$

So the absolute error in taking (X' – Y') as an approximation to (X – Y) is less than or equal to the sum of the absolute errors in taking X' as approximation to X and Y' is an approximation to Y.

(c) Error in multiplication operation

In order to find the absolute error E_A in the multiplication of two numbers X and Y as given through

$E_A = (X + E_{AX}) (Y + E_{AY}) - XY$

where E_{AX} and E_{AY} are the absolute errors in X and Y respectively.

$E_A = X E_{AY} + Y E_{AX} + E_{AX} E_{AY}$

Assume $E_{AX} E_{AY} << 0$, then

$E_A = X E_{AY} + Y E_{AX}$

(d) Error in division operation

Similarly the absolute error E_A in the quotient oftwo numbers X and Y as given through

$$E_A = \frac{X + E_{AX}}{Y + E_{AY}} - \frac{X}{Y} = \frac{Y\,E_{AX} + X\,E_{AY}}{Y\,(Y + E_{AY})}$$

$$E_A = \frac{Y\,E_{AX} + X\,E_{AY}}{Y^2\,(1 + \frac{E_{AY}}{Y})}$$

Assume $\frac{E_{AY}}{Y} << 0$, then

$$E_A = \frac{Y\,E_{AX} + X\,E_{AY}}{Y^2}$$

$$E_A = \frac{X}{Y}\left(\frac{E_{AX}}{X} - \frac{E_{AY}}{Y}\right)$$

1.8 Error Estimation

The difference between the true value and measured (or approximate) value is termed as total error, which consists of truncation error and round-off error. The purpose of numerical method is to decrease the error as much possible as and obtain the solution with least errors, so error analysis / estimation becomes important. The relative error of an approximate solution is the ratio of total error to the exact value. It is of greater significance than error itself for the true value becomes larger, then a larger error may be acceptable. If the true value decreases, then the error must also decreases otherwise the computed results may be absurd.

Example 1.10: Estimate the error in computing the volume of sphere, whose radius is r with an error of ±0.01 cm. The volume is function of radius as, $f(r) = \frac{4}{3}\pi r^3$.

Solution: $f'(r) = 4\pi r^2$

If r_t is the true value and r_m is the measured value, then

$-0.01 \le (r_t - r_m) \le 0.01$

As

$f(r_t) - f(r_m) \approx f'(r_t)\,(r_t - r_m)$

So the error in volume measurement is

$-0.01\, f'(r_t) \le f(r_t) - f(r_m) \le 0.01\, f'(r_t)$

Suppose r_t is 10 cm, then the error bounds for the volume measurement will be

$-4\pi \leq f(r_t) - f(r_m) \leq 4\pi$

As $f'(r_t) = 4\pi\ (10)^2$. Similarly say 0.05% error was made in measurement of radius then

$$\frac{r_t - r_m}{r_t} X 100 = 0.05$$

Hence the corresponding % error estimation of volume is

$$\frac{f(r_t) - f(r_m)}{f(r_t)} X 100 \approx \frac{f(r_t) - f(r\text{-}r_t)}{f(r_t)} X 100$$

$$= \frac{4\pi r_t^2\ (r\text{-}r_t)}{\frac{4}{3}\pi r_t^3} X 100 = \frac{3\ (r\text{-}r_t)}{r_t} X 100 = 3 X 0.05 = 0.15$$

1.9 General Error Formula

In order to find the error through truncated or rounded off of certain number (i.e., true value, having n decimal digits) to a certain number (i.e., measured or approximate value), the following rules are used.

(a) Absolute error (E_A) due to truncation upto k digits

$|E_A| < 10^{n-k}$

|True value – Measured value| < 10^{n-k}

Example 1.11: Find the absolute error of number 0.00123456, which is truncated to 3 decimal digits.

Solution: True value = 0.00123856 = $0.123456 * 10^{-2}$

After truncated, measured or approximate value = $0.123 * 10^{-2} = 0.123 * 10^{n}$

when, n = -2, k = 3

(E_A) Absolute error = |True value – Measured value|
= $*0.000856 * 10^{-2} = 0.856 * 10^{-5}$

$10^{n-k} = 10^{-2-3} = 10^{-5}$

So, $E_A < 10^{n-k}$

(b) Relative error (E_R) due to truncation upto k digits

$$\text{Relative error } (E_R) = \frac{\text{True value - Measured value}}{\text{True value}} = \frac{E_A}{\text{True value}}$$

$$\left|\frac{E_A}{\text{True value}}\right| < 10^{1-k}$$

Example 1.12: Find the relative error of number 0.004997, which is truncated to 3 decimal digits.

Solution: True value = 0.004997 = 0.4997 * 10^{-2}

After truncated, measured or approximate value = 0.499 * 10^{-2}= 0.499 * 10^{n}

n = -2, k = 3

$$\left|\frac{E_A}{\text{True value}}\right| = \left|\frac{0.4997*10^{-2} - 0.499*10^{-2}}{0.4997*10^{-2}}\right| = 0.140 * 10^{-2}$$

$10^{1-k} = 10^{1-3} = 10^{-2}$

So, $E_R < 10^{1-k}$

(c) Absolute error (E_A) due to rounding off to k digits

$|E_A| < 0.5*10^{n-k}$

|True value – Measured value| < $0.5*10^{n-k}$

Example 1.13: Find the absolute error of number 0.00123456, which is rounded off to 3 decimal digits.

Solution: True value = 0.00123856 = 0.123856 * 10^{-2}

After rounded off, measured or approximate value = 0.124 * 10^{-2}= 0.124 * 10^{n}

when, n = -2, k = 3

(E_A) Absolute error = |True value – Measured value|

= *0.000144*10^{-2} = 0.144*10^{-3}*10^{-2}

= 0.144 * 10^{-5}

$10^{n-k} = 10^{-2-3} = 10^{-5}$

So, $E_A < 0.5 * 10^{n-k}$

(d) Relative error (E_R) due to truncation upto k digits

$$\text{Relative error } (E_R) = \frac{\text{True value - Measured value}}{\text{True value}} = \frac{E_A}{\text{True value}}$$

$$\left|\frac{E_A}{\text{True value}}\right| < 0.5*10^{1-k}$$

Example 1.14: Find the relative error of number 0.004997, which is rounded off to 3 decimal digits.

Solution: True value = 0.004997 = 0.4997 * 10^{-2}

After rounded off, measured or approximate value = 0.500 * 10^{-2} = 0.500 * 10^{n}

n = -2, k = 3

$$\left|\frac{E_A}{\text{True value}}\right| = \left|\frac{0.4997*10^{-2} - 0.500*10^{-2}}{0.4997*10^{-2}}\right| = 0.600 * 10^{-3}$$

$10^{1-k} = 10^{1-3} = 10^{-2}$

So, $E_R < 0.5 * 10^{1-k}$

If a number is correct to n significant digits, then the maximum relative error ≤ 0.5 * 10^{-n} and if a number is correct upto d decimal places, then the absolute error ≤ 0.5 * 10^{-d}

If significant figure of number is k and the number is correct to n significant figures, then relative error < $\dfrac{1}{k * 10^{n-1}}$

1.9.1 Function Approximation

Suppose y = f(x_1, x_2) be a function of two variables x and x_2. If δx_1 and δx_2 are the errors in x_1 and x_2 respectively, then the error δy in y is

$$y + \delta y = f(x_1 + \delta x_1, x_2 + \delta x_2)$$

Upon expansion the right side of equation through Taylor's series, yields

$$y + \delta y = f(x_1, x_2) = \left(\frac{\partial f}{\partial x_1}\delta x_1 + \frac{\partial f}{\partial x_2}\delta x_2\right) + \text{terms with higher powers of } \delta x_1 \text{ and } \delta x_2$$

If errors δx_1 and δx_2 are so small and hence their squares and higher powers are neglected, yields

$$\delta y = \frac{\partial f}{\partial x_1}\delta x_1 + \frac{\partial f}{\partial x_2}\delta x_2 \text{ approx.}$$

Generally, the error δy in the function y = f(x_1, x_2..... x_n) corresponding to the errors δx_i in x_i, where i = 1, 2.... n; is given through

$$\delta y \approx \frac{\partial f}{\partial x_1}\delta x_1 + \frac{\partial f}{\partial x_2}\delta x_2 + ... + \frac{\partial f}{\partial x_n}\delta x_n$$

and the relative error in y is $E_R = \dfrac{\delta y}{y} \approx \dfrac{\partial f}{\partial x_1}\dfrac{\delta x_1}{y} + \dfrac{\partial f}{\partial x_2}\dfrac{\delta x_2}{y} + ... + \dfrac{\partial f}{\partial x_n}\dfrac{\delta x_n}{y}$

Example 1.15: If u = $\frac{4x^2y^y}{z^4}$ and errors in x, y, z is 0.001, compute the relative maximum error in u when x = y = z = 1.

Solution: $\frac{\partial u}{\partial x}=\frac{8xy^3}{z^4}, \frac{\partial u}{\partial y}=\frac{12x^2y^2}{z^4}, \frac{\partial u}{\partial z}=-\frac{16x^2y^3}{z^4}$

As the errors δx, δy, δz may be +ve or –ve, so consider the absolute value of right side terms

$$(\delta u)_{\text{maximum}} = \left|\frac{8xy^3}{z^4}\delta x\right| + \left|\frac{12x^2y^2}{z^4}\delta y\right| + \left|\frac{16x^2y^3}{z^4}\delta z\right|$$

$= (8 * 0.001) + (12 * 0.001) + (16 * 0.001) = 0.036$

1.9.2 Stability and Condition

Any numerical method is said to be convergent when the approximate solution approaches the exact solution as h tends to zero, provided the rounding off errors appears from the initial conditions approaches zero (horizontal axis function is divided into n sub-intervals each of width or step size 'h'). This means that a method is continuously refined by taking smaller and smaller step sizes and hence the sequence of approximate solutions must converge to the exact solution. There is limit to which 'h', which can be decreased for controlling the truncation error, beyond which a further decrease in h will results in increase in round off error, ultimately increase of total error.

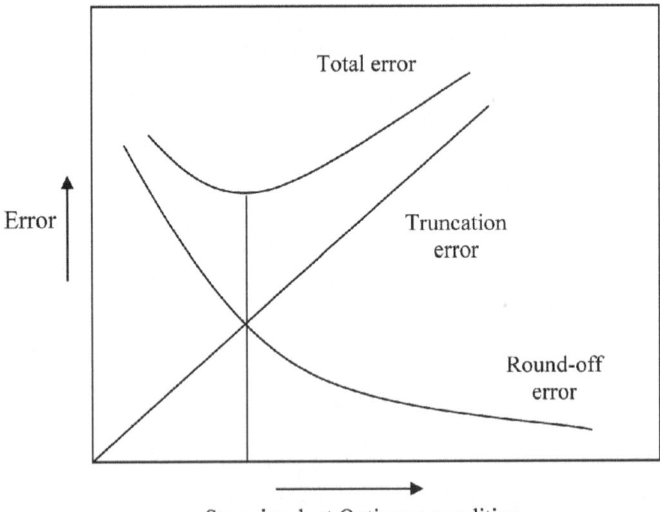

Fig. 1.2: Graph (error vs step size) represents type of errors

So in order to deal with usage of stable methods can be done. A numerical method is called **stable** if it produces a bounded solution which initiates the exact solution and it is **unstable** if the rounded-off errors at one stage / step of the computation process will propagate with increase in magnitude in further stages / steps. A mathematical problem is said to be **unconditionally (or absolutely) stable** if a method is stable for all values of the parameters and is **conditionally stable** as the method is stable for some values of the parameters i.e., a small change in a problem data results in small changes in the solution. For example, in case of numerical methods for ordinary differential equations, the Taylor's method and Adams-Bashforth method are stable, while Euler's method and Runge-Kutta method are conditionally stable.

1.9.3 Uncertainty in Data or Noise

In the physical data, the uncertainty or error is based on which the computation model is based can generate errors during analysis, and this type of error is called **noise**. The data affects the numerical computation's accuracy. The error or noise produces inaccuracy and imprecision. If the input data having 'y' significant digits of accuracy then the results through computation must be in 'y' significant digits of accuracy.

Refer *Section 1.7.4*

Exercise

1.1 Compare various methods of problem solving.

1.2 Give the significance of numerical methods.

1.3 Explain Taylor's theorem. Solve $y' = x^2 + y$, $y(0) = 1$. Find values of y at $x = 0.1$.

1.4 Write a note on number system.

1.5 Conversion into number system

(a) Conversion into decimal number system from binary number 1011, octal number 2983 and hexadecimal number 97AF

(b) Decimal number is 35, convert into binary number, octal number and hexadecimal number

1.6 Perform the operations

(a) $(2)_8 + (10)_8$

(b) Convert $(4D3)_{16}$ to base 10

1.7 Determine the absolute and relative errors involved if $x=2/3$ is represented in normalized decimal form with 6 digits by round-off and truncation.

1.8 If $x = 2.537$, then determine the absolute and relative error when

(a) x is rounded,

(b) x is truncated to 2 decimal places

1.9 Explain errors and its types.

1.10 Show that the relative error E_r of the product where $x = x_e + \Delta x$ and $y = y_e + \Delta y$ is $E_r = E_x + E_y$. Assume $|E_x| << 1$ and $|E_y| << 1$.

1.11 If h = and errors in x, y, z is 0.01, then calculate the relative maximum error in h when $x = y = z = 1$.

1.12 Give the condition of stability.

2

Linear System of Equations

2.1 Introduction

Consider two linear equations in two variables i.e., x and y, such that equations are

$$2x - 3y = 4$$

$$3x + y = 1$$

These equations are linear equations because of presence of terms x, y & some numbers, and there are no terms having x^2 and y^2, or x*y, or any higher powers of x and y. So in order to find the solution of these equations, the unknown numbers x and y have to satisfy both the equations. Hence, system or pair of equations is called **simultaneous equations**. So the solution is x = 1 and y = –2, which satisfies both the linear equations.

Simultaneous linear equations occur often in science and engineering. For example, these are used in finding the cost of any processing / chemical plant or chemical reaction, output of processing / chemical plant, analysis of electronic circuits including invariant elements, analysis of network under certain conditions etc.

2.1.1 Introduction to Determinants

The expression $\begin{vmatrix} a_{11} & a_{12} \\ a_{21} & a_{22} \end{vmatrix}$ is called a **determinant** of the 2nd order. It contains 4 elements (a_{11}, a_{12}, a_{21}, a_{22}) and represents [$a_{11}\,a_{22} - a_{12}\,a_{21}$]. Here 1st subscript of a_{11} represents row i.e., 1st row and 2nd subscript represents column i.e., 1st column. There are 2 rows and 2 columns in 2nd order determinant. The diagonal from top corner left to right bottom corner (containing elements i.e., a_{11} and a_{22}) is called **leading diagonal**.

Similarly, $\begin{vmatrix} a_{11} & a_{12} & a_{13} \\ a_{21} & a_{22} & a_{23} \\ a_{31} & a_{32} & a_{33} \end{vmatrix}$ is 3rd order determinant and contains 9 elements arranged in 3 rows and 3 columns.

Cofactor in determinant: It is the determinant obtained through deleting the row and the column that intersect at that element with the proper sign and is represented by capital letter.

sign = $(-1)^{i+j}$ where i is the row and j is the column

For example, cofactor of a_{32} is $A_{32} = (-1)^{3+2} \begin{vmatrix} a_{11} & a_{13} \\ a_{21} & a_{23} \end{vmatrix} = -1\ [a_{11}\, a_{23} - a_{13}\, a_{21}]$.

This can be done by freezing the row and column of the corresponding element and write all other elements (as shown above for element a_{32}) and subtract the product of diagonal opposites.

A **determinant can be expanded** in terms of any row or column by multiplying each element of the row / column in terms of which we intend expanding the determinants, by its cofactor and then add up all these products. This expression is represented by 'Δ'. For example, expand row 1, yields

$$\Delta = a_{11}\, A_{11} + a_{12}\, A_{12} + a_{13}\, A_{13}$$

$$\Delta = a_{11} \begin{vmatrix} a_{22} & a_{23} \\ a_{32} & a_{33} \end{vmatrix} - a_{12} \begin{vmatrix} a_{21} & a_{23} \\ a_{31} & a_{33} \end{vmatrix} + a_{13} \begin{vmatrix} a_{21} & a_{22} \\ a_{31} & a_{32} \end{vmatrix}$$

$$\Delta = a_{11}\,(a_{22}\, a_{33} - a_{32}\, a_{23}) - a_{12}\,(a_{21}\, a_{33} - a_{31}\, a_{23}) + a_{13}\,(a_{21}\, a_{32} - a_{31}\, a_{22})$$

Properties of determinants

(1) A determinant remains same on changing rows and columns.

(2) On changing two parallel lines (rows or columns), the numerical value remains same but with opposite sign.

(3) If two parallel lines (rows or columns) are identical, then determinant is zero.

(4) If each element of a line be multiplied with same factor resulting in multiplication of same factor with determinant.

(5) If each element of line is added equi-multiples of the corresponding elements of one or more parallel lines, the determinant remain same. For example

$$\begin{vmatrix} a_{11}+xa_{12}-ya_{13} & a_{12} & a_{13} \\ a_{21}+xa_{22}-ya_{23} & a_{22} & a_{23} \\ a_{31}+xa_{32}-ya_{33} & a_{32} & a_{33} \end{vmatrix}$$

$$= \begin{vmatrix} a_{11} & a_{12} & a_{13} \\ a_{21} & a_{22} & a_{23} \\ a_{31} & a_{32} & a_{33} \end{vmatrix} + \mathrm{x} \begin{vmatrix} a_{12} & a_{12} & a_{13} \\ a_{22} & a_{22} & a_{23} \\ a_{32} & a_{32} & a_{33} \end{vmatrix} - y \begin{vmatrix} a_{13} & a_{12} & a_{13} \\ a_{23} & a_{22} & a_{23} \\ a_{33} & a_{32} & a_{33} \end{vmatrix} = \Delta + 0 + 0 = \Delta$$

Multiplication of Determinants

$$\begin{vmatrix} a_{11} & a_{12} & a_{13} \\ a_{21} & a_{22} & a_{23} \\ a_{31} & a_{32} & a_{33} \end{vmatrix} X \begin{vmatrix} b_{11} & b_{12} & b_{13} \\ b_{21} & b_{22} & b_{23} \\ b_{31} & b_{32} & b_{33} \end{vmatrix} =$$

$$\begin{vmatrix} a_{11}b_{11}+a_{12}b_{12}+a_{13}b_{13} & a_{21}b_{11}+a_{22}b_{12}+a_{13}b_{13} & a_{31}b_{11}+a_{32}b_{12}+a_{33}b_{13} \\ a_{11}b_{21}+a_{12}b_{22}+a_{13}b_{23} & a_{21}b_{21}+a_{22}b_{22}+a_{23}b_{23} & a_{31}b_{11}+a_{32}b_{22}+a_{33}b_{23} \\ a_{11}b_{31}+a_{12}b_{23}+a_{13}b_{33} & a_{21}b_{31}+a_{22}b_{23}+a_{33}b_{33} & a_{31}b_{11}+a_{32}b_{32}+a_{33}b3 \end{vmatrix}$$

2.1.2 Introduction to Matrices

Matrices are everywhere and used in day-today's life to make presentation of numbers clearer and make calculations easier to program. In case of spreadsheet such as Excel or written numbers in a table, matrix of numbers are used. A matrix mn numbers arranged in a rectangular array of m rows and n columns is called mXn matrix and is represented by

$$\mathrm{A}=\begin{bmatrix} a_{11} & a_{12} & \cdots & a_{1n} \\ a_{21} & a_{22} & \cdots & a_{2n} \\ \vdots & \vdots & \vdots & \vdots \\ a_{m1} & a_{m2} & \cdots & a_{mn} \end{bmatrix} = [\mathrm{a}_{ij}]$$

Points related to matrices

(1) **Row matrix**: A matrix with single row, such as $\begin{bmatrix} a_{11} & a_{12} & \cdots & a_{1n} \end{bmatrix}$

(2) **Column matrix**: A matrix with single column, such as $\begin{bmatrix} a_{11} \\ a_{21} \\ \vdots \\ a_{m1} \end{bmatrix}$

(3) **Vector:** It is a matrix that has only one row or one column. There are two types of vectors i.e., row vectors and column vectors.

(A) **Row Vector:** If a matrix [B] has one row, it is called a row vector and is the dimension of the row vector. $[B] = [b_1\ b_2\ \ldots\ldots b_n]$

(B) **Column vector:** If a matrix [C] has one column, it is called a column vector $[C] = \begin{bmatrix} c_1 \\ \vdots \\ \vdots \\ c_m \end{bmatrix}$ and is the dimension of the vector.

(4) **Submatrix:** If some row(s) or/and column(s) of a matrix [A] are deleted (no rows or columns may be deleted), the remaining matrix is called a submatrix of [A]. For example, consider matrix, $[A] = \begin{bmatrix} 4 & 6 & 2 \\ 3 & -1 & 2 \end{bmatrix}$ and corresponding to this matrix, submatices are

$$\begin{bmatrix} 4 & 6 & 2 \\ 3 & -1 & 2 \end{bmatrix}, \begin{bmatrix} 4 & 6 \\ 3 & -1 \end{bmatrix}, [4\ \ 6\ \ 2], [4], \begin{bmatrix} 2 \\ 2 \end{bmatrix}$$

(5) **Square matrix**: A matrix with equal number of rows and columns, such as $[a_{ij}]$, where i = j; $\begin{bmatrix} a_{11} & a_{12} & a_{13} \\ a_{21} & a_{22} & a_{23} \\ a_{31} & a_{32} & a_{33} \end{bmatrix}$ having 3 rows and 3 columns. A square matrix is said to be singular if determinant is zero, otherwise non-singular. The diagonal elements are called **leading diagonal** and their sum is called **trace of the matrix**.

(6) **Unit matrix**: If diagonal elements are unity for matrix of order n, such as $\begin{bmatrix} 1 & a_{12} & a_{13} \\ a_{21} & 1 & a_{23} \\ a_{31} & a_{32} & 1 \end{bmatrix}$

(7) **Null matrix**: If all the elements of the matrix are zero.

(8) **Symmetric matrix**: When $a_{ij} = a_{ji}$ for all the vales of i and j for square matrix $[a_{ij}]$, such as $\begin{bmatrix} a & h & g \\ h & b & f \\ g & f & c \end{bmatrix}$

(9) **Skew-symmetric matrix**: When $a_{ij} = -a_{ji}$ for all the vales of i and j for square matrix $[a_{ij}]$, such as $\begin{bmatrix} 0 & h & -g \\ -h & 0 & f \\ g & -f & 0 \end{bmatrix}$

(10) **Upper Traingular matrix**: For a square matrix, all elements which are below the leading diagonal are zero, such as $\begin{bmatrix} 1 & u_{12} & u_{13} \\ 0 & 1 & u_{23} \\ 0 & 0 & 1 \end{bmatrix}$

(11) **Lower Traingular matrix**: For a square matrix, whose elements which are above the leading diagonal are zero, such as $\begin{bmatrix} l_{11} & 0 & 0 \\ l_{21} & l_{22} & 0 \\ l_{31} & l_{32} & l_{33} \end{bmatrix}$

(12) **Diagonal matrix:** A square matrix with all non-diagonal elements equal to zero is called a diagonal matrix, that is, only the diagonal entries of the square matrix can be non-zero, ($a_{ij} = 0, i \neq j$). For example,

$$[A] = \begin{bmatrix} 3 & 0 & 0 \\ 0 & 2.1 & 0 \\ 0 & 0 & 5 \end{bmatrix}$$

(13) **Zero matrix:** A matrix whose all entries are zero is called a zero matrix,

($a_{ij} = 0$ for all i and j). For example, $[A] = \begin{bmatrix} 0 & 0 & 0 \\ 0 & 0 & 0 \\ 0 & 0 & 0 \end{bmatrix}$

(14) **Tridiagonal matrices:** It is a square matrix in which all elements not on the following are zero - the major diagonal, the diagonal above the major diagonal, and the diagonal below the major diagonal. For example,

$$[A] = \begin{bmatrix} 2 & 4 & 0 & 0 \\ 2 & 3 & 9 & 0 \\ 0 & 0 & 5 & 2 \\ 0 & 0 & 3 & 6 \end{bmatrix}$$

Operations on matrices

(1) **Addition and subtraction:** For equality of matrices, they must be of same order and each element of matrix 1 is equal to the corresponding element of matrix 2. Consider two matrices, A and B of order 2, then resulting matrix is

Addition matrix, C = A + B

Subtraction matrix, D = A – B

(2) **Multiplication:** Two matrices can be multiplied only when the number of columns in the 1st matrix is equal to the number of rows in the 2nd matrix. For example, if A and B are two matrices or order (m*n) and (n*p) respectively, then their product C=AB is of matrix (m*p).

$$\begin{bmatrix} a_{11} & a_{12} & a_{13} \\ a_{21} & a_{22} & a_{23} \\ a_{31} & a_{32} & a_{33} \\ a_{41} & a_{42} & a_{43} \end{bmatrix} X \begin{bmatrix} b_{11} & b_{12} \\ b_{21} & b_{22} \\ b_{31} & b_{23} \end{bmatrix} = \begin{bmatrix} a_{11}b_{11} + a_{12}b_{21} + a_{13}b_{31} & a_{11}b_{12} + a_{12}b_{22} + a_{13}b_{32} \\ a_{21}b_{11} + a_{22}b_{21} + a_{23}b_{31} & a_{21}b_{12} + a_{22}b_{22} + a_{23}b_{32} \\ a_{31}b_{11} + a_{32}b_{21} + a_{33}b_{31} & a_{31}b_{12} + a_{32}b_{12} + a_{33}b_{32} \\ a_{41}b_{11} + a_{42}b_{21} + a_{43}b_{31} & a_{41}b_{12} + a_{42}b_{12} + a_{43}b_{32} \end{bmatrix}$$

Here matrix A is of order (4*3) and B is of order (3*2), then the resulting matrix C is of order (4*2).

If matrix A is multiplied with scalar k, then matrix with elements are k times the corresponding elements of matrix A.

(3) **Transpose of Matrix:** The matrix obtained through changing rows and columns. It is represented by A'.

Consider matrix $A = \begin{bmatrix} a_1 & b_1 & c_1 \\ a_2 & b_2 & c_2 \\ a_3 & b_3 & c_3 \end{bmatrix}$

Transpose of matrix $A = A' = \begin{bmatrix} a_1 & a_2 & a_3 \\ b_1 & b_2 & b_3 \\ c_1 & c_2 & c_3 \end{bmatrix}$

Some noteworthy points are

(i) A' = A, for symmetric matrix

(ii) A' = –A, for skew-symmetric matrix

(iii) (AB)' = B' A'

(iv) For square matrix A i.e., A = ½ (A + A') + ½ (A – A') = B + C

where B is a symmetric matrix; B'= ½ (A + A')' = ½ (A' + A) = B, and

C is a skew-symmetric matrix; C = ½ (A – A')' = ½ (A' – A) = –C

(4) **Adjoint of a Square Matrix:** For a square matrix A, the transported matrix of cofactors A and is represented as adj A.

Explained in section 2.3

(5) **Inverse of a Matrix:** For A is a non-singular nth order square matrix, then B square matrix of same order can be written as AB = BA = I, where I is unit matrix.

Inverse of A = A^{-1} = adj/|A|

$A A^{-1} = A^{-1} A = I$

Explained in section 2.3

(6) **Rank of a Matrix:** In a given matrix A, selection of rth row and r^{th} column and then deletion of all other rows and columns, then the determinant by these r*r elements is called minor of order r of matrix A. Hence there will be a number of different minors of the same order, yielded through deleting different rows and columns from the same matrix. A matrix is said to be rank r when (i) matrix has atleast one non-zero minor of rth order, (ii) every minor of order higher than r vanishes.

(7) **Elementary transformations of a matrix**: The operations, (i) interchange of any two rows (or columns); (ii) multiplication of any two rows (or

columns) with a non-zero number; (iii) addition of constant multiple of the elements of any roe (or column) to the corresponding elements of any other row (or column).

(8) Equivalent matrix: Consider two matrices A and B, which are equivalent if one can be obtained from the other through elementary transformations sequence. It is represented by '– '.

(9) Elementary matrix: This matrix is obtained through unit matrix through any elementary transformations.

(10) Normal form of a matrix: Every non-zero matrix A of rth order, can be reduced through a sequence of elementary transformations, to form $\begin{bmatrix} T_r & 0 \\ 0 & 0 \end{bmatrix}$ matrix.

2.2 Methods of Solution

The solution of these linear system of equations can be obtained through methods i.e., direct or iterative methods.

(1) Direct methods

Direct methods yields the solution after a certain amount of fixed computations. Such methods are

(a) Gauss elimination method,

(b) Gauss Jordan method.

Gauss elimination method is time consuming method and it requires more of recording, but Gauss Jordan method is quite time saving as compare to Gauss elimination method.

(2) Iterative methods

Through iterative methods, the solution of system of linear equations starts with an approximation and leads to an exact solution. Iterative methods depends upon the number of iterations, more the iteration more accurate is the solution. Iterative methods are given preference over direct methods because of lesser round-off errors and faster & self correcting process (i.e., error at 1st stage is automatically corrected in 2nd stage). Iterative methods may not always converge i.e., may not provide solution always, but on convergence these methods are given preference over direct methods. Such methods are:

(a) Jacobi's iteration method,

(b) Gauss-Seidal iteration method,

(c) Crout's method,

(d) Cholesky's traingularisation method.

Among these iterative methods, Crout's method is given preference for solving linear system of equation and through software computation.

2.3 The Inverse of a Matrix

In matrix operation, there is no division operation, so for the denominator, we have to find the inverse of matrix and then multiply with the numerator.

Consider matrix $A = \begin{bmatrix} a_1 & b_1 & c_1 \\ a_2 & b_2 & c_2 \\ a_3 & b_3 & c_3 \end{bmatrix}$

Then find the determinant through expansion of any row / column, suppose expand row 1 as

$|A| = \Delta = a_1 (b_2 c_3 - b_3 c_2) - b_1 (a_2 c_3 - a_3 c_2) + c_1 (a_2 b_3 - a_3 b_2)$

If $\Delta = 0$, then the matrix has no inverse.

Transpose of matrix $A = A' = \begin{bmatrix} a_1 & a_2 & a_3 \\ b_1 & b_2 & b_3 \\ c_1 & c_2 & c_3 \end{bmatrix}$ i.e., changing rows and columns

Then calculate the determinants of minors of the elements $a_1, a_2, a_3 \ldots c_3$ can be written as $A_1, A_2, A_3 \ldots C_3$ as $\begin{bmatrix} A_1 & A_2 & A_3 \\ B_1 & B_2 & B_3 \\ C_1 & C_2 & C_3 \end{bmatrix}$

For element a_1, $A_1 = \begin{vmatrix} b_2 & b_3 \\ c_2 & c_3 \end{vmatrix} = (b_2 c_3 - b_3 c_2)$;

For element a_2, $A_2 = \begin{vmatrix} b_1 & b_3 \\ c_1 & c_3 \end{vmatrix} = (b_1 c_3 - b_3 c_1)$;

For element a_3, $A_3 = \begin{vmatrix} b_1 & b_2 \\ c_1 & c_2 \end{vmatrix} = (b_1 c_2 - b_2 c_1)$

Similarly $B_1 = \begin{vmatrix} a_2 & a_3 \\ c_2 & c_3 \end{vmatrix} = (a_2c_3 - a_3c_2)$; $B_2 = \begin{vmatrix} a_1 & a_3 \\ c_1 & c_3 \end{vmatrix} = (a_1c_1 - a_3c_1)$; $B_3 = \begin{vmatrix} a_1 & a_2 \\ c_1 & c_2 \end{vmatrix}$ $= (a_1c_2 - a_2c_1)$

$C_1 = \begin{vmatrix} a_2 & a_3 \\ b_2 & b_3 \end{vmatrix} = (a_2b_3 - a_3b_2)$; $C_2 = \begin{vmatrix} a_1 & a_3 \\ b_1 & b_3 \end{vmatrix} = (a_1b_3 - a_3b_1)$; $C_3 = \begin{vmatrix} a_1 & a_2 \\ b_1 & b_2 \end{vmatrix} = (a_1b_2$ $- a_2b_1)$

Create the matrix of cofactors by placing the results into a new matrix of by aligning each minor matrix determinant with the corresponding position in the original matrix. While assigning the sign, keeping the 1st element sign same and then alternate sign i.e., first (+), then (–)

i.e., $\begin{bmatrix} + & - & + \\ - & + & - \\ + & - & + \end{bmatrix}$

Calculate Adjoint A, adj A $= \begin{bmatrix} A_1 & A_2 & A_3 \\ B_1 & B_2 & B_3 \\ C_1 & C_2 & C_3 \end{bmatrix} X \begin{bmatrix} + & - & + \\ - & + & - \\ + & - & + \end{bmatrix} = \begin{bmatrix} A_1 & -A_2 & A_3 \\ -B_1 & B_2 & -B_3 \\ C_1 & -C_2 & C_3 \end{bmatrix}$

Hence the Inverse of matrix A i.e., $A^{-1} = \frac{1}{|A|} adj\ A = \frac{1}{|A|} \begin{bmatrix} A_1 & -A_2 & A_3 \\ -B_1 & B_2 & -B_3 \\ C_1 & -C_2 & C_3 \end{bmatrix}$

Example 2.1: Find the inverse of the matrix A, $\begin{bmatrix} 1 & 2 & 3 \\ 0 & 1 & 4 \\ 5 & 6 & 0 \end{bmatrix}$

Solution: The given matrix $A = \begin{bmatrix} 1 & 2 & 3 \\ 0 & 1 & 4 \\ 5 & 6 & 0 \end{bmatrix}$

Determinant of matrix A is, Δ, $|A| = 1\ (0 - 24) - 2\ (0 - 20) + 3\ (0 - 5) = 1$

Transpose of matrix A = A' = $\begin{bmatrix} 1 & 0 & 5 \\ 2 & 1 & 6 \\ 3 & 4 & 0 \end{bmatrix}$ i.e., changing rows and columns

Calculation of minors:

$$A_1 = \begin{vmatrix} 1 & 6 \\ 4 & 0 \end{vmatrix} = -24;\ A_2 = \begin{vmatrix} 2 & 6 \\ 3 & 0 \end{vmatrix} = -18;\ A_3 = \begin{vmatrix} 2 & 1 \\ 3 & 4 \end{vmatrix} = 5$$

$$B_1 = \begin{vmatrix} 0 & 5 \\ 4 & 0 \end{vmatrix} = -20;\ B_2 = \begin{vmatrix} 1 & 6 \\ 3 & 0 \end{vmatrix} = -15;\ B_3 = \begin{vmatrix} 1 & 0 \\ 3 & 4 \end{vmatrix} = 4$$

$$C_1 = \begin{vmatrix} 0 & 5 \\ 1 & 6 \end{vmatrix} = -5;\ C_2 = \begin{vmatrix} 1 & 6 \\ 2 & 6 \end{vmatrix} = -4;\ C_3 = \begin{vmatrix} 1 & 0 \\ 2 & 1 \end{vmatrix} = 1$$

Calculate Adjoint A, adj A = $\begin{bmatrix} A_1 & A_2 & A_3 \\ B_1 & B_2 & B_3 \\ C_1 & C_2 & C_3 \end{bmatrix} X \begin{bmatrix} + & - & + \\ - & + & - \\ + & - & + \end{bmatrix}$

$$= \begin{bmatrix} -24 & -18 & 5 \\ -20 & -15 & 4 \\ -5 & -4 & 1 \end{bmatrix} X \begin{bmatrix} + & - & + \\ - & + & - \\ + & - & + \end{bmatrix} = \begin{bmatrix} -24 & 18 & 5 \\ 20 & -15 & -4 \\ -5 & 4 & 1 \end{bmatrix}$$

Inverse of matrix A i.e., $A^{-1} = \frac{1}{|A|} adj\ A = \frac{1}{1} \begin{bmatrix} -24 & 18 & 5 \\ 20 & -15 & -4 \\ -5 & 4 & 1 \end{bmatrix}$

Hence the inverse of the matrix A is $A^{-1} = \begin{bmatrix} -24 & 18 & 5 \\ 20 & -15 & -4 \\ -5 & 4 & 1 \end{bmatrix}$

2.4 Cramer's Rule

It is explicit formula regarding the solution of system of linear equations with number of equations as that of number of variables, provided

(a) for square system, A X = D, will have unique solution for every column matrix D, if determinant of square matrix A, $\Delta \neq 0$.

(b) for a homogeneous square system, A X = 0, will have trivial solution x = 0, if $\Delta \neq 0$.

This method is based on the determinants of square matrix and of the matrices that are obtained through replacement of one column with the column vector D i.e., on the right hand side of equation (A X = D). This rule was given by Gabriel Cramer in 1750.

Procedure: Consider the system of equation

$$a_1 x + b_1 y + c_1 z = d_1 \ldots\ldots\ldots\ldots\ldots\ldots (1)$$

$$a_2 x + b_2 y + c_2 z = d_2 \ldots\ldots\ldots\ldots\ldots\ldots (2)$$

$$a_3 x + b_3 y + c_3 z = d_3 \ldots\ldots\ldots\ldots\ldots\ldots (3)$$

$$\text{A X = D, where A} = \begin{bmatrix} a_1 & b_1 & c_1 \\ a_2 & b_2 & c_2 \\ a_3 & b_3 & c_3 \end{bmatrix}, X = \begin{bmatrix} x \\ y \\ z \end{bmatrix}, D = \begin{bmatrix} d_1 \\ d_2 \\ d_3 \end{bmatrix}$$

If the determinant of coefficients is $\Delta = \begin{vmatrix} a_1 & b_1 & c_1 \\ a_2 & b_2 & c_2 \\ a_3 & b_3 & c_3 \end{vmatrix}$

Then, $x\,\Delta = \begin{vmatrix} x a_1 & b_1 & c_1 \\ x a_2 & b_2 & c_2 \\ x a_3 & b_3 & c_3 \end{vmatrix}$

Operate $C_1 + yC_2 + zC_3$ yields, $\begin{vmatrix} a_1x + b_1y + c_1z & b_1 & c_1 \\ a_2x + b_2y + c_2z & b_2 & c_2 \\ a_3x + b_3y + c_3z & b_3 & c_3 \end{vmatrix} = \begin{vmatrix} d_1 & b_1 & c_1 \\ d_2 & b_2 & c_2 \\ d_3 & b_3 & c_3 \end{vmatrix} = D_x$

From system of equations (1), (2) and (3),
So,

$$x = D_x / \Delta = \begin{vmatrix} d_1 & b_1 & c_1 \\ d_2 & b_2 & c_2 \\ d_3 & b_3 & c_3 \end{vmatrix} / \begin{vmatrix} a_1 & b_1 & c_1 \\ a_2 & b_2 & c_2 \\ a_3 & b_3 & c_3 \end{vmatrix} \text{ provided } \Delta \neq 0, \quad (4)$$

Similarly,

$$y = D_y / \Delta = \begin{vmatrix} a_1 & d_1 & c_1 \\ a_2 & d_2 & c_2 \\ a_3 & d_3 & c_3 \end{vmatrix} / \begin{vmatrix} a_1 & b_1 & c_1 \\ a_2 & b_2 & c_2 \\ a_3 & b_3 & c_3 \end{vmatrix} \tag{5}$$

$$z = D_z / \Delta = \begin{vmatrix} a_1 & b_1 & d_1 \\ a_2 & b_2 & d_2 \\ a_3 & b_3 & d_3 \end{vmatrix} / \begin{vmatrix} a_1 & b_1 & c_1 \\ a_2 & b_2 & c_2 \\ a_3 & b_3 & c_3 \end{vmatrix} \tag{6}$$

So the equations (4), (5) and (6) provides the values of unknown variables x, y and z. This rule fails when $\Delta = 0$ and this method requires too much calculations i.e., for system of 10 equations this method need $7*10^7$ calculations rather than other method, resulting in consumption of more time, energy and money.

Example 2.2: Solve the system of equations through Cramer's rule

$2x + y + z = 3$

$x - y - z = 0$

$x + 2y + z = 0$

Solution: The given system of equations

$$2x + y + z = 3 \tag{1}$$

$$x - y - z = 0 \tag{2}$$

$$x + 2y + z = 0 \tag{3}$$

$$A\,X = D \tag{4}$$

$$\text{where } A = \begin{bmatrix} a_1 & b_1 & c_1 \\ a_2 & b_2 & c_2 \\ a_3 & b_3 & c_3 \end{bmatrix} = \begin{bmatrix} 2 & 1 & 1 \\ 1 & -1 & -1 \\ 1 & 2 & 1 \end{bmatrix}; X = \begin{bmatrix} x \\ y \\ z \end{bmatrix}; D = \begin{bmatrix} d_1 \\ d_2 \\ d_3 \end{bmatrix} = \begin{bmatrix} 3 \\ 0 \\ 0 \end{bmatrix}$$

$$\text{If the determinant of coefficients is } \Delta = \begin{vmatrix} a_1 & b_1 & c_1 \\ a_2 & b_2 & c_2 \\ a_3 & b_3 & c_3 \end{vmatrix} = \begin{vmatrix} 2 & 1 & 1 \\ 1 & -1 & -1 \\ 1 & 2 & 1 \end{vmatrix}$$

On expansion of row 1 as;

$$\Delta = 2\,(-1 + 2) - 1\,(1 + 1) + 1\,(2 + 1) = 3$$

Replace column 1 of Δ with D as

$$D_x = \begin{vmatrix} d_1 & b_1 & c_1 \\ d_2 & b_2 & c_2 \\ d_3 & b_3 & c_3 \end{vmatrix} = \begin{vmatrix} 3 & 1 & 1 \\ 0 & -1 & -1 \\ 0 & 2 & 1 \end{vmatrix} = 3\ (-1 + 2) - 1\ (0) + 1\ (0) = 3;$$

[on expansion of row 1]

Similarly

Replace column 2 of Δwith D as

$$D_y = \begin{vmatrix} a_1 & d_1 & c_1 \\ a_2 & d_2 & c_2 \\ a_3 & d_3 & c_3 \end{vmatrix} = \begin{vmatrix} 2 & 3 & 1 \\ 1 & 0 & -1 \\ 1 & 0 & 1 \end{vmatrix} = 2\ (0) - 3\ (1 + 1) + 1\ (0) = -6;$$

[on expansion of row 1]

Replace column 3 of Δ with D as

$$D_z = \begin{vmatrix} a_1 & b_1 & d_1 \\ a_2 & b_2 & d_2 \\ a_3 & b_3 & d_3 \end{vmatrix} = \begin{vmatrix} 2 & 1 & 3 \\ 1 & -1 & 0 \\ 1 & 2 & 0 \end{vmatrix} = 2\ (0) - 1\ (0) + 3(2 + 1) = 9;$$

[on expansion of row 1]

So,

$x = D_x / \Delta = 3 / 3 = 1$

$y = D_y / \Delta = -6 / 3 = -2$

$z = D_z / \Delta = 9 / 3 = 3$

Hence the solution of system of equations; $x = 1$, $y = -2$, $z = 3$.

2.5 Matrix Inversion Method

In this general method, the given matrix is written in AX=D form.

Given matrix:
$$\begin{aligned} a_1 x + b_1 y + c_1 z &= d_1 \\ a_2 x + b_2 y + c_2 z &= d_2 \\ a_3 x + b_3 y + c_3 z &= d_3 \end{aligned}$$

$$\text{where } A = \begin{bmatrix} a_1 & b_1 & c_1 \\ a_2 & b_2 & c_2 \\ a_3 & b_3 & c_3 \end{bmatrix}, X = \begin{bmatrix} x \\ y \\ z \end{bmatrix} \text{and } D = \begin{bmatrix} d_1 \\ d_2 \\ d_3 \end{bmatrix}$$

Upon multiplication of inverse of given matrix on both sides yields solution of unknown variables (X) of system of linear equations. But this method does not work when the given matrix is a singular matrix i.e., $|A|=0$ and also not suitable for large system of equations hence the method becomes bulky and consumes a lot of time, energy and money because of evaluation of A^{-1} by cofactors.

Procedure: Consider the system of equation

$$\text{If } A = \begin{bmatrix} a_1 & b_1 & c_1 \\ a_2 & b_2 & c_2 \\ a_3 & b_3 & c_3 \end{bmatrix}, X = \begin{bmatrix} x \\ y \\ z \end{bmatrix} \text{and } D = \begin{bmatrix} d_1 \\ d_2 \\ d_3 \end{bmatrix}$$

Then system of equations can be written as,

$$AX = D \tag{4}$$

Multiply inverse of A i.e., A^{-1} on both sides of equation (4),

$$A^{-1}AX = A^{-1}D \tag{5}$$

As, $A^{-1}A = I$, then, $IX = A^{-1}D$

$$X = A^{-1}D \tag{6}$$

$$\begin{bmatrix} x \\ y \\ z \end{bmatrix} = \frac{1}{|A|}\begin{bmatrix} A_1 & A_2 & A_3 \\ B_1 & B_2 & B_3 \\ C_1 & C_2 & C_3 \end{bmatrix} X \begin{bmatrix} d_1 \\ d_2 \\ d_3 \end{bmatrix}$$

where A_1, B_1,.... etc are the cofactors of a_1, a_2,.... etc in the determinant $|A|$

Example 2.3: Apply matrix inversion method to solve the system of equations

$x + 2y = 4$

$3x - 5y = 1$

Solution: In $AX = D$

$$\text{where } A = \begin{bmatrix} a_1 & b_1 \\ a_2 & b_2 \end{bmatrix} = \begin{bmatrix} 1 & 2 \\ 3 & -5 \end{bmatrix}, X = \begin{bmatrix} x \\ y \end{bmatrix}, D = \begin{bmatrix} d_1 \\ d_2 \end{bmatrix} = \begin{bmatrix} 4 \\ 1 \end{bmatrix}$$

$|A| = (1)(-5) - (2)(3) = -11$

$$\text{adj A} = \text{adj}\begin{bmatrix} 1 & 2 \\ 3 & -5 \end{bmatrix} = \begin{bmatrix} -5 & -2 \\ -3 & 1 \end{bmatrix}$$

$$X = A^{-1} D$$

$$\begin{bmatrix} x \\ y \end{bmatrix} = \frac{adj\ A}{|A|} X \begin{bmatrix} d_1 \\ d_2 \end{bmatrix}$$

$$\begin{bmatrix} x \\ y \end{bmatrix} = \frac{1}{-11}\begin{bmatrix} -5 & -2 \\ -3 & 1 \end{bmatrix} X \begin{bmatrix} 4 \\ 1 \end{bmatrix}$$

$$\begin{bmatrix} x \\ y \end{bmatrix} = \frac{1}{-11}\begin{bmatrix} -22 \\ -11 \end{bmatrix} = \begin{bmatrix} 2 \\ 1 \end{bmatrix}$$

Hence x = 2, y = 1 is the solution of the given system of equations.

Example 2.4: Apply matrix inversion method to solve the system of equations

3x + y + 2z = 3

2x – 3y – z = –3

x + 2y + z = 4

Solution: In A X = D

$$\text{where A} = \begin{bmatrix} a_1 & b_1 & c_1 \\ a_2 & b_2 & c_2 \\ a_3 & b_3 & c_3 \end{bmatrix} = \begin{bmatrix} 3 & 1 & 2 \\ 2 & -3 & -1 \\ 1 & 2 & 1 \end{bmatrix}, X = \begin{bmatrix} x \\ y \\ z \end{bmatrix}, D = \begin{bmatrix} d_1 \\ d_2 \\ d_3 \end{bmatrix} = \begin{bmatrix} 3 \\ -3 \\ 4 \end{bmatrix}$$

Determinant of matrix A is, Δ,

$$|A| = \begin{vmatrix} 3 & 1 & 2 \\ 2 & -3 & -1 \\ 1 & 2 & 1 \end{vmatrix} = 3\ (-3 + 2) - 1\ (2 + 1) + 2\ (4 + 3) = 8$$

[expansion of row 1]

$$\text{Transpose of matrix A i.e., } A' = \begin{bmatrix} 3 & 2 & 1 \\ 1 & -3 & 2 \\ 2 & -1 & 1 \end{bmatrix}$$

Calculation of minors:

$$A_1 \begin{vmatrix} -3 & 2 \\ -1 & 1 \end{vmatrix} = -1;\ A_2 = \begin{vmatrix} 1 & 2 \\ 2 & 1 \end{vmatrix} = -3;\ A_3 = \begin{vmatrix} 1 & -3 \\ 2 & -1 \end{vmatrix} = 5$$

$$B_1 = \begin{vmatrix} 2 & 1 \\ -1 & 1 \end{vmatrix} = 3;\ B_2 = \begin{vmatrix} 3 & 1 \\ 2 & 1 \end{vmatrix} = 1;\ B_3 = \begin{vmatrix} 3 & 2 \\ 2 & -1 \end{vmatrix} = -7$$

$$C_1 = \begin{vmatrix} 2 & 1 \\ -3 & 2 \end{vmatrix} = 7;\ C_2 = \begin{vmatrix} 3 & 1 \\ 1 & 2 \end{vmatrix} = 5;\ C_3 = \begin{vmatrix} 3 & 2 \\ 1 & -3 \end{vmatrix} = -11$$

$$\text{Adjoint A, adj A} = \begin{bmatrix} A_1 & A_2 & A_3 \\ B_1 & B_2 & B_3 \\ C_1 & C_2 & C_3 \end{bmatrix} = \begin{bmatrix} + & - & + \\ - & + & - \\ + & - & + \end{bmatrix}$$

$$= \begin{bmatrix} -1 & -3 & 5 \\ 3 & 1 & -7 \\ 7 & 5 & -11 \end{bmatrix} X \begin{bmatrix} + & - & + \\ - & + & - \\ + & - & + \end{bmatrix} = \begin{bmatrix} -1 & 3 & 5 \\ -3 & 1 & 7 \\ 7 & -5 & -11 \end{bmatrix}$$

$$\text{Inverse of matrix A i.e., } A^{-1} = \frac{1}{|A|} adj\ A = \frac{1}{8} \begin{bmatrix} -1 & 3 & 5 \\ -3 & 1 & 7 \\ 7 & -5 & -11 \end{bmatrix}$$

$X = A^{-1} D$

$$\begin{bmatrix} x \\ y \\ z \end{bmatrix} = \frac{adj\ A}{|A|} X \begin{bmatrix} d_1 \\ d_2 \\ d_3 \end{bmatrix}$$

$$\begin{bmatrix} x \\ y \\ z \end{bmatrix} = \frac{1}{8} \begin{bmatrix} -1 & 3 & 5 \\ -3 & 1 & 7 \\ 7 & -5 & -11 \end{bmatrix} X \begin{bmatrix} 3 \\ -3 \\ 4 \end{bmatrix}$$

$$\begin{bmatrix} x \\ y \\ z \end{bmatrix} = \begin{bmatrix} 1 \\ 2 \\ 1 \end{bmatrix}$$

Hence x = 1, y = 2, z = 1 is the solution of the given system of equations.

2.5.1 Augmented Matrix

An augmented matrix is a matrix obtained by appending the columns of two given matrices, usually for performing the same elementary row operations on each of the given matrices.

Consider the given matrix is

$x - 2y + 3z = 7$

$2x + y + z = 4$

$-3x + 2y - 2z = -10$

The augmented matrix is

$$\left[\begin{array}{ccc|c} 1 & -2 & 3 & 7 \\ 2 & 1 & 1 & 4 \\ -3 & 2 & -2 & -10 \end{array}\right]$$

The first row consists of all the constants from the first equation with the coefficient of the x in the first column, the coefficient of the y in the second column, the coefficient of the z in the third column and the constant in the final column. The second row is the constants from the second equation with the same placement and likewise for the third row. The solid line (or dashed line) represents where the equal sign was in the original system of equations and is not always included.

Elementary row operations

(1) Interchange two rows: In this operation we will interchange all the entries in row i and row j. The notation is $R_i \leftrightarrow R_j$ i.e., $R_1 \leftrightarrow R_3$

$$\left[\begin{array}{ccc|c} 1 & -2 & 3 & 7 \\ 2 & 1 & 1 & 4 \\ -3 & 2 & -2 & -10 \end{array}\right] \rightarrow \left[\begin{array}{ccc|c} -3 & 2 & -2 & -10 \\ 2 & 1 & 1 & 4 \\ 1 & -2 & 3 & 7 \end{array}\right]$$

Every entry in the third row moves up to the first row and every entry in the first row moves down to the third row.

(2) Multiply / divide a row by a constant c: Multiply the ith row with constant c, i.e., cR_i and for division the notation is $(1/c)R_i$. Operate $-4R_3 \leftrightarrow R_3$

$$\left[\begin{array}{ccc|c} 1 & -2 & 3 & 7 \\ 2 & 1 & 1 & 4 \\ -3 & 2 & -2 & -10 \end{array}\right] \rightarrow \left[\begin{array}{ccc|c} 1 & -2 & 3 & 7 \\ 2 & 1 & 1 & 4 \\ -12 & -8 & 8 & 40 \end{array}\right]$$

(3) Multiply of a row with another row: Replace the ith row with the sum of row i and constant c times j^{th} row i.e., $R_3 - 4R_1 \rightarrow R_3$

$$\left[\begin{array}{ccc|c} 1 & -2 & 3 & 7 \\ 2 & 1 & 1 & 4 \\ -3 & 2 & -2 & -10 \end{array}\right] \rightarrow \left[\begin{array}{ccc|c} 1 & -2 & 3 & 7 \\ 2 & 1 & 1 & 4 \\ -7 & 10 & -14 & -38 \end{array}\right]$$

Perform elementary row operations in such to convert it into $\left[\begin{array}{ccc|c} 1 & 0 & 0 & m \\ 0 & 1 & 0 & n \\ 0 & 0 & 1 & o \end{array}\right]$ to find the solution i.e., x = m, y = n, z = o.

Suppose ax + by = p and cx + dy = q are two matrices and hence the augmented matrix is

$$\left[\begin{array}{cc|c} a & b & p \\ c & d & q \end{array}\right]$$

Perform elementary row operations in such to convert it into $\left[\begin{array}{cc|c} 1 & 0 & h \\ 0 & 1 & k \end{array}\right]$. Hence the solution is x = h, y = k.

This method is called **Gauss-Jordan Elimination**.

Example 2.5: Solve the system of equations

3x – 2y = 14

x + 3y = 1

Solution: The given system of equations

3x – 2y = 14

x + 3y = 1

The augmented matrix in the form = $\left[\begin{array}{cc|c} a & b & p \\ c & d & q \end{array}\right] = \left[\begin{array}{cc|c} 3 & -2 & 14 \\ 1 & 3 & 1 \end{array}\right]$

Convert the element 'a' into '1' by interchanging R_1 and R_3 i.e., $R_1 \leftrightarrow R_3$;

$$\left[\begin{array}{cc|c} 1 & 3 & 1 \\ 3 & -2 & 14 \end{array}\right]$$

Convert the element 'c' into '0' through operation $R_2 - 3R_1 \rightarrow R_2$; $\left[\begin{array}{cc|c} 1 & 3 & 1 \\ 0 & -11 & 11 \end{array}\right]$

Convert the element 'd' into '1' through operation (-1/11) $R_2 \rightarrow R_2$; $\left[\begin{array}{cc|c} 1 & 3 & 1 \\ 0 & 1 & -1 \end{array}\right]$

Convert the element 'b' into '0' through operation $R_1 - 3R_2 \rightarrow R_1$; $\left[\begin{array}{cc|c} 1 & 0 & 4 \\ 0 & 1 & -1 \end{array}\right]$

Therefore the final augmented matrix has solution; x = 4 and y = –1.

2.6 Gauss Elimination Method

In this method, from given linear system of equations the unknowns are eliminated successively and the system is transformed into an equivalent system in upper triangular form; this form can be solved easily through back substitution. This is very simple method and adopted for computer operations.

Procedure: Consider a linear system having three equations

$$a_{11}\, x_1 + a_{12}\, x_2 = a_{13}\, x_3 = b_1 \qquad (1)$$

$$a_{21}\, x_2 + a_{22}\, x_2 = a_{23}\, x_3 = b_2 \qquad (2)$$

$$a_{31}\, x_1 + a_{32}\, x_2 = a_{33}\, x_3 = b_3 \qquad (3)$$

(a) Upper triangularization

1st stage: Eliminate the coefficients a_{21} and a_{31} through equation 1. To achieve this

Operate [(equation 2) – $\left(\frac{a_{21}}{a_{11}}\right)$ (equation 1)], results, and

$$a'_{22}\, x_2 + a'_{23}\, x_3 = b'_2 \qquad (4)$$

Operate [(equation 3) – $\left(\frac{a_{31}}{a_{11}}\right)$ (equation 1)], results, $a'_{32}\, x_2 + a'_{33}\, x_3 = b'_3$ (5)

So the new system of equations is

$$a_{11}\, x_1 + a_{12}\, x_2 = a_{13}\, x_3 = b_1 \qquad (1)$$

$$a'_{22}\, x_2 + a'_{23}\, x_3 = b'_2 \qquad (4)$$

$$a'_{32}\, x_2 + a'_{33}\, x_3 = b'_3 \qquad (5)$$

The numbers $\frac{a_{21}}{a_{11}}$ and $\frac{a_{31}}{a_{11}}$ are called **multipliers** for 1st stage and it is assumed that $a_{11} \neq 0$. a_{11} is the **1st pivot** for the 1st stage and 1st equation is called **pivotal equation**.

2nd stage: Eliminate the coefficientsfrom equation 5. To achieve this

Operate [(equation 5) – $\left(\frac{a'_{32}}{a'_{22}}\right)$ (equation 4)], results, $a''_{33}\, x''_3 = b''_3$ (6)

So the new system of equations is

$a_{11}\, x_1 + a_{12}\, x_2 = a_{13}\, x_3 = b_1$ (1)

$a'_{22}\, x_2 + a'_{23}\, x_3 = b'_2$ (4)

$a''_{32}\, x_2 + a_{33}\, x_3 = b''_3$ (6)

The number $\frac{a'_{32}}{a'_{22}}$ is called **multipliers** for 2nd stage and it is assumed that $a'_{22} \neq O$. a'_{22} is the **2nd pivot** for the 2nd stage and 2nd equation is called **2nd pivotal equation**.

(b) Back substitution: From equations 6, results

$$x_3 = \frac{b''_3}{a''_{33}}$$

Put the value ofin equation 4, yields, $x_2 = \frac{b''_3}{a'_{22}}(b'_2 - a'_{23}\, x_3)$

Put the value ofandin equation 1, yields, $x_1 = \frac{1}{a_{11}}(b_1 - a_{12}\, x_2 - a_{13}\, x_3)$

This method is most used on computers and it is clear that the pivot elements i.e.,, a_{11}, a'_{22} and a''_{33} are used as divisors in multipliers in order to find the solution of linear equation and these must be non-zero. If they are zero, then need to re-arrange the equation and if not possible then system of equations has no solution.

If a pivot element is small as compared to column's elements then the corresponding multiplier will be greater than 1. As large multipliers through elimination and back substitution process yields high round-off errors. If pivot is zero, then again this method stops working. So in order to avoid round-off errors and getting zero pivot element, re-arranging of remaining rows are required, called **pivoting**. Pivoting is of two types

(i) Partial pivoting: In step 1, the largest numerical coefficient of x is chosen from all the equations and brought as 1st pivot by interchanging the 1st equation with the equation having largest coefficient of x. In the step 2, the largest numerical coefficient of y is chosen from the remaining equations (except equation 1) and brought as 2nd pivot by interchanging the 2nd equation with the equation having largest coefficient of y. This procedure is followed till equation with the single variable is reached.

(ii) Complete pivoting: This is more complicated procedure but yields high accuracy. If specific order for eliminating x, y and z is not required then selection of numerically largest coefficient from the entire matrix of coefficients can be one. This requires not only interchange of equations but also an interchange of the position of the variables.

Example 2.6: 'without pivoting'; Solve the system of equations with the application of Gauss elimination method.

$3x + 4y + 5z = 40,$

$2x - 3y + 4z = 13,$

$x + y + z = 9$

Solution: The given system of equations is

$$3x + 4y + 5z = 40 \quad (1)$$

$$2x - 3y + 4z = 13 \quad (2)$$

$$x + y + z = 9 \quad (3)$$

(a) Upper triangularization

1st stage: Eliminate x from equation (2) and (3).

Multiply the equation (1) with $\frac{2}{3}$ and subtracts from equation (2)

$$-\frac{17}{3}y + \frac{2}{3}z = -\frac{41}{3} \quad (2.a)$$

Multiply the equation (1) with $\frac{1}{3}$ and subtracts from equation (3)

$$-\frac{1}{3}y - \frac{2}{3}z = -\frac{13}{3} \quad (3.a)$$

The system of equations becomes

$$3x + 4y + 5z = 40 \quad (1)$$

$$-\frac{17}{3}y + \frac{2}{3}z = -\frac{41}{3} \qquad (2.a)$$

$$-\frac{1}{3}y - \frac{2}{3}z = -\frac{13}{3} \qquad (3.a)$$

2nd stage: Eliminate y from equation (3.a)

Multiply equation (2.a) with $\frac{1}{17}$ and subtract from equation (3.a)

$$\frac{12}{17}z = -\frac{180}{51} \qquad (3.b)$$

The system of equations becomes

$$3x + 4y + 5z = 40 \qquad (1)$$

$$-\frac{17}{3}y + \frac{2}{3}z = -\frac{41}{3} \qquad (2.a)$$

$$\frac{12}{17}z = -\frac{180}{51} \qquad (3.b)$$

(b) Back substitution

From equation (3.b), z = 5.

From equation (2.a), upon putting the value of z yields y; y = 3.

From equation (1), upon putting the value of y and z which yields x; x = 1.

Hence the solution of given system of equations is given by

x = 1, y = 3 and z = 5

Example 2.7: 'with pivoting'; Solve the system of equations i.e.,

3x + 6y + z = 16

2x + 4y + 3z = 13

x + 3y + 2z = 9

with the application of Gauss elimination method.

Solution: The given system of equations is

$$3x + 6y + z = 16 \qquad (1)$$

$$2x + 4y + 3z = 13 \qquad (2)$$

$$x + 3y + 2z = 9 \qquad (3)$$

(a) Upper triangularization

1st stage: Eliminate x from equation (2) and (3).

Multiply the equation (1) with $\frac{2}{3}$ and subtracts from equation (2)

$$\frac{7}{3}z = \frac{7}{3} \qquad (2.a)$$

Multiply the equation (1) with $\frac{1}{3}$ and subtracts from equation (3)

$$y + \frac{5}{3}z = \frac{11}{3} \qquad (3.a)$$

The system of equations becomes

$$3x + 6y + z = 16 \qquad (1)$$

$$\frac{7}{3}z = \frac{7}{3} \qquad (2.a)$$

$$y + \frac{5}{3}z = \frac{11}{3} \qquad (3.a)$$

In equation 2.a, the pivot of variable y is zero, so it is not possible to proceed further. There is need to interchange the equations 2.a and 3.a. So the system of equation becomes upper triangular form as

$$3x + 6y + z = 16 \qquad (1)$$

$$y + \frac{5}{3}z = \frac{11}{3} \qquad (2.a)$$

$$\frac{7}{3}z = \frac{7}{3} \qquad (3.a)$$

(b) Back substitution

From equation (3.a), z = 1.

From equation (2.a), upon putting the value of z yields y; y = 2.

From equation (1), upon putting the value of y and z which yields x; x = 1.

Hence the solution of given system of equations is given by

x = 1, y = 2 and z = 1

Example 2.8: 'with partial pivoting'; Solve the system of equations i.e.,

$3a + b + c - 2d = -10$

$4a + 0\,b + 2c + d = 8$

$3a + 2b + 2c + 0\,d = 7$

$a + 3b + 2c - d = -5$

with the application of Gauss elimination method.

Solution: The given system of equations is

$3a + b + c - 2d = -10$ (1)

$4a + 0\,b + 2c + d = 8$ (2)

$3a + 2b + 2c + 0\,d = 7$ (3)

$a + 3b + 2c - d = -5$ (4)

(a) Upper triangularization

1st stage: Interchange 1st and 2nd equation in order to make the 1st pivot element largest in its column. The new system of equations becomes

$4a + 0\,b + 2c + d = 8$ (1)

$3a + b + c - 2d = -10$ (2)

$3a + 2b + 2c + 0\,d = 7$ (3)

$a + 3b + 2c - d = -5$ (4)

Eliminate a from equation (2), (3) and (4).

Multiply the equation (1) with $\frac{1}{2}$ and subtracts from equation (2) and equation (2) becomes

$b + 0\,c - \frac{5}{2}d = -14$ (2.a)

Multiply the equation (1) with $\frac{3}{4}$ and subtracts from equation (3) and equation (3) becomes

$2b + \frac{1}{2}c - \frac{3}{4}d = 1$ (3.a)

Multiply the equation (1) with $\frac{1}{4}$ and subtracts from equation (4) and equation (4) becomes

$$3b + \frac{3}{2}c - \frac{5}{4}d = -7 \quad (4.a)$$

The system of equations becomes

$$4a + 0\,b + 2c + d = 8 \quad (1)$$

$$b + 0\,c - \frac{5}{2}d = -14 \quad (2.a)$$

$$2b + \frac{1}{2}c - \frac{3}{4}d = 1 \quad (3.a)$$

$$3b + \frac{3}{2}c - \frac{5}{4}d = -7 \quad (4.a)$$

2nd stage: Interchange (2.a) and (4.a) equation in order to make the 2nd pivot element largest in its column. The new system of equations becomes

$$4a + 0\,b + 2c + d = 8 \quad (1)$$

$$3b + \frac{3}{2}c - \frac{5}{4}d = -7 \quad (2.a)$$

$$2b + \frac{1}{2}c - \frac{3}{4}d = 1 \quad (3.a)$$

$$b + 0\,c - \frac{5}{2}d = -14 \quad (4.a)$$

Eliminate b from equation (3.a) and (4.a).

Multiply the equation (2.a) with $\frac{2}{3}$ and subtracts from equation (3.a) and equation (3.a) becomes

$$-\frac{1}{2}c + \frac{1}{12}d = \frac{17}{3} \quad (3.b)$$

Multiply the equation (2.a) with $\frac{1}{3}$ and subtracts from equation (4.a) and equation (4.a) becomes

$$-\frac{1}{2}c-\frac{25}{12}d=-\frac{35}{3} \qquad (4.b)$$

The system of equations becomes

$$4a+0\,b+2c+d=8 \qquad (1)$$

$$3b+\frac{3}{2}\,c-\frac{5}{4}\,d=-7 \qquad (2.a)$$

$$-\frac{1}{2}c+\frac{1}{12}d=\frac{17}{3} \qquad (3.b)$$

$$-\frac{1}{2}c-\frac{25}{12}d=-\frac{35}{3} \qquad (4.b)$$

3rd stage: The coefficients of equation (3.b) and (4.b) are same, so no need of interchanging equations.

Eliminate c from equation (4.b).

Subtracts equation (3.b) from equation (4.b) and equation (4.b) becomes

$$-\frac{13}{6}d=-\frac{52}{3} \qquad (4.c)$$

The system of equations becomes

$$4a+0\,b+2c+d=8 \qquad (1)$$

$$3b+\frac{3}{2}c-\frac{5}{4}d=-7 \qquad (2.a)$$

$$-\frac{1}{2}c+\frac{1}{12}d=\frac{17}{3} \qquad (3.b)$$

$$-\frac{13}{6}d=-\frac{52}{3} \qquad (4.c)$$

(b) Back substitution

From equation (4.c), d = 8.

From equation (3.b), upon putting the value of d yields c; c = – 10.

From equation (2.a), upon putting the value of c and d which yields b; b = 6.

From equation (1), upon putting the value of a, b and c, which yields a; a = 5.

Hence the solution of given system of equations is given by

a = 5, b = 6, c = –10 and d = 8

2.7 Gauss-Jordan Method

This method is a little modified form of Gauss elimination method. In this method, at all stages the elimination of coefficients is done not only below the pivot elements but also in the equations above pivot elements, which ultimately the system of equations are reduced to diagonal matrix form.

Advantage is reduction in back substitution in order to find the unknown variables and hence this method becomes cost and time effective as compared to Gauss elimination method. This method is also suitable for implementation through computer.

Disadvantage is large amount of multiplications to be performed to get the values of unknown variables as compared to Gauss elimination method. Consider a system of 10 equations; Gauss-Jordan method requires ~500 arithmetic operations i.e. $(n^2/2)$, whereas Gauss elimination method requires 333 i.e., $(n^3/2)$. So Gauss-Jordan method appears to be easy but with number of multiplications. In view point of this, Gauss elimination method is still given preference.

Procedure: Consider a linear system having three equations

$$a_{11}\, x_1 + a_{12}\, x_2 + a_{13}\, x_3 = b_1 \quad (1)$$

$$a_{21}\, x_2 + a_{22}\, x_2 + a_{23}\, x_3 = b_2 \quad (2)$$

$$a_{31}\, x_1 + a_{32}\, x_2 + a_{33}\, x_3 = b_3 \quad (3)$$

(a) 1st stage: Eliminate the coefficients and through equation 1.1. To achieve this

Operate $\left(\dfrac{\text{equation 1}}{a_{11}}\right)$, results, $x_1 + \dfrac{a_{12}}{a_{11}} x_2 + \dfrac{a_{13}}{a_{11}} x_3 = \dfrac{b_1}{a_{11}}$ or

$$x_1 + a'_{12}\, x_2 + a'_{13}\, x_3 = b'_1 \quad (1.1)$$

Operate [(equation 2) – (a_{21}) (equation 1.1)], results,

$$0\, x_1 + a'_{22}\, x_2 + a'_{23}\, x_3 = b'_2 \quad (2.1) \text{ and}$$

Operate [(equation 3) – (a_{31}) (equation 1.1)], results,

$$0\, x_1 + a'_{32}\, x_2 + a'_{33}\, x_3 = b'_3 \quad (3.1)$$

So the new system of equations is

$$x_1 + a'_{12}\, x_2 + a'_{13}\, x_3 = b'_1 \quad (1.1)$$

$$0\, x_1 + a'_{22}\, x_2 + a'_{23}\, x_3 = b'_2 \quad (2.1)$$

$$0\, x_1 + a'_{32}\, x_2 + a'_{33}\, x_3 = b'_3 \quad (3.1)$$

(b) 2nd stage: Eliminate the coefficients a'_{12} and a'_{32} through equation 5. To achieve this

Operate $\left(\frac{\text{equation 2.1}}{a'_{22}}\right)$, results in change of equation 5 to, $0\,x_1 + x_2 + \frac{a'_{23}}{a'_{22}} x_3 = \frac{b'_3}{a'_{22}}$

or

$$0\ x_1 + x_2 + a''_{23} x_3 = b''_2 \tag{2.2}$$

Operate [(equation 1.1) – (a'_{12}) (equation 2.2)], results, and

$$x_1 + 0\ x_2 + a''_{13} x_3 = b''_1. \tag{1.2}$$

Operate [(equation 3.1) – (a'_{32}) (equation 2.2)], results,

$$0\ x_1 + 0\ x_2 + a''_{33} x_3 = b''_3. \tag{3.2}$$

So the new system of equations is

$$x_1 + 0\ x_2 + a''_{13} x_3 = b''_1 \tag{2.2}$$

$$0\ x_1 + x_2 + a''_{23} x_3 = b''_2 \tag{1.2}$$

$$0\ x_1 + 0\ x_2 + a''_{33} x_3 = b''_3. \tag{3.2}$$

(c) 3nd stage: Eliminate the coefficients a''_{13} and a''_{23} through equation 9. To achieve this

Operate $\left(\frac{\text{equation 3.2}}{a''_{33}}\right)$, results in change of equation 3.2 to, $0\,x_1 + 0\,x_2 + x_3 = \frac{b'_3}{a'_{33}}$

or

$$0x_1 + 0x_2 + x_3 = b'''_3 \tag{3.3}$$

Operate [(equation 1.2) – $\left(a''_{13}\right)$ (equation 3.3)], results,

$x_1 + 0\ x_2 + 0\ x_3 = b'''_1$.................. (1.3) and

Operate [(equation 2.2) – (a''_{12}) (equation 3.3)], results,

$$0\ x_1 + x_2 + 0\ x_3 = b'''_2 \tag{2.3}$$

So the new system of equations is

$$x_1 + 0\ x_2 + 0\ x_3 = b'''_1 \tag{1.3}$$

$$0\ x_1 + x_2 + 0\ x_3 = b'''_2. \tag{2.3}$$

$$0x_1 + 0x_2 + x_3 = b'''_3 \tag{3.3}$$

So from the view point of equations 1.3, 2.3 & 3.3, the solution of linear system i.e., finding unknown variables, is given through numbers on the right side of the equations.

Example 2.9: Solve the system of equations 2x + 4y + 2z = 16; 2x + 3y+ 4z = 20; 4x + 3y + 2z = 16 with the application of Gauss Jordan method.

Solution: The given system of equations is

$$2x + 4y + 2z = 16 \tag{1}$$

$$2x + 3y + 4z = 20 \tag{2}$$

$$4x + 3y + 2z = 16 \tag{3}$$

(a) 1st stage: Eliminate the coefficients of x from 2nd and 3rd equation through 1st equation. To achieve this,

Operate $\left(\dfrac{\text{equation 1}}{2}\right)$, results, $x + 2y + z = 8$ (1.1)

Operate [(equation 2) – (2) (equation 1.1)], results, $0\,x - y + 2z = 4$ (2.1)

Operate [(equation 3) – (4) (equation 1.1)], results, $0\,x - 5y - 2z = -16$ (3.1)

So the new system of equations is

$$x + 2y + z = 8 \tag{1.1}$$

$$0\,x - y + 2z = 4 \tag{2.1}$$

$$0\,x - 5y - 2z = -16 \tag{3.1}$$

(b) 2nd stage: Eliminate the coefficients of y from 1.1 and 3.1 equation through 2.1 equation. To achieve this,

Operate $\left(\dfrac{\text{equation 2.1}}{-1}\right)$, results, $0\,x + y - 2z = -4$ (2.2)

Operate [(equation 1.1) – (2) (equation 2.2)], results, $x + 0\,y + 5z = 16$ (1.2)

Operate [(equation 3.1) – (–5) (equation 2.2)], results, $0\,x + 0\,y - 12z = -36$ (3.2)

So the new system of equations is

$$x + 0\,y + 5z = 16 \tag{1.2}$$

$$0\,x + y - 2z = -4 \tag{2.2}$$

$$0\,x + 0\,y - 12z = -36 \tag{3.2}$$

(c) 3rd stage: Eliminate the coefficients of z from 1.2 and 2.2 equation through 3.2 equation. To achieve this,

Operate $\left(\frac{\text{equation 3.2}}{-12}\right)$, results, 0 x + 0 y + z = 3 (3.3)

Operate [(equation 1.2) – (5) (equation 3.3)], results, x + 0 y + 0 z = 1 (1.3)

Operate [(equation 2.2) – (–2) (equation 3.3)], results, 0 x + y + 0 z = 2 (2.3)

So the new system of equations is

x + 0 y + 0 z = 1 (1.3)

0 x + y + 0 z = 2 (2.3)

0 x + 0 y + z = 3 (3.3)

From equations 1.3, 2.3 & 2.3, the values of variables are x = 1, y = 2 and z = 3.

2.8 Crout's Method

This method is also called '**LU Decomposition method**' or **factorization method**' or '**Doolittle's method**'. In this method, every square matrix is a product of a lower triangular matrix and upper triangular matrix, provided all the principal minors of square matrix are non-singular. If A = [a_{ij}], then

$$a_{11} \neq 0, \begin{bmatrix} a_{11} & a_{12} \\ a_{21} & a_{22} \end{bmatrix} \neq 0, \begin{bmatrix} a_{11} & a_{12} & a_{13} \\ a_{21} & a_{22} & a_{23} \\ a_{31} & a_{32} & a_{33} \end{bmatrix} \neq \text{ 0, and so on.}$$

Consider the system of equations

$$a_{11}\, x_1 + a_{12}\, x_2 + a_{13}\, x_3 = b_1$$

$$a_{11}\, x_1 + a_{12}\, x_2 + a_{13}\, x_3 = b_2$$

$$a_{11}\, x_1 + a_{12}\, x_2 + a_{13}\, x_3 = b_3$$

So write this system of equations in the form **AX = B** as

$$\text{in which, A} = \begin{bmatrix} a_{11} & a_{12} & a_{13} \\ a_{21} & a_{22} & a_{23} \\ a_{31} & a_{32} & a_{33} \end{bmatrix}, X = \begin{bmatrix} x_1 \\ x_2 \\ x_3 \end{bmatrix} and\ B = \begin{bmatrix} b_1 \\ b_2 \\ b_3 \end{bmatrix}$$

Let, **A = LU**

where L and U are the lower and upper triangular matrix respectively.

$$L = \begin{bmatrix} l_{11} & 0 & 0 \\ l_{21} & l_{22} & 0 \\ l_{31} & l_{32} & l_{33} \end{bmatrix} \text{ and } U = \begin{bmatrix} 1 & u_{12} & u_{13} \\ 0 & 1 & u_{23} \\ 0 & 0 & 1 \end{bmatrix}$$

A = L U

$$\begin{bmatrix} a_{11} & a_{12} & a_{13} \\ a_{21} & a_{22} & a_{23} \\ a_{31} & a_{32} & a_{33} \end{bmatrix} = \begin{bmatrix} l_{11} & 0 & 0 \\ l_{21} & l_{22} & 0 \\ l_{31} & l_{32} & l_{33} \end{bmatrix} \begin{bmatrix} 1 & u_{12} & u_{13} \\ 0 & 1 & u_{23} \\ 0 & 0 & 1 \end{bmatrix}$$

Multiply the matrix on the right hand side,

$$\begin{bmatrix} a_{11} & a_{12} & a_{13} \\ a_{21} & a_{22} & a_{23} \\ a_{31} & a_{32} & a_{33} \end{bmatrix} = \begin{bmatrix} l_{11} & l_{11}u_{12} & l_{11}u_{13} \\ l_{21} & l_{21}u_{12}+l_{22} & l_{21}u_{13}+l_{22}u_{23} \\ l_{31} & l_{31}u_{12}+l_{32} & l_{31}u_{13}+l_{32}u_{23}+l_{33} \end{bmatrix}$$

Comparing and equating corresponding elements in the matrix of both sides as

(a) Compare 1st column

$$l_{11} = a_{11}, l_{21} = a_{21} l_{31} = a_{311}$$

(b) Compare 1st row and put above values

$$l_{11}\ u_{12} = a_{12} \Rightarrow u_{12} = \frac{a_{12}}{l_{11}} = \frac{a_{12}}{a_{11}}$$

$$l_{11}\ u_{13} = a_{13} \Rightarrow u_{13} = \frac{a_{13}}{l_{11}} = \frac{a_{13}}{a_{11}}$$

(c) Compare 2nd column

$$l_{21}u_{12} + l_{22} = a_{22} \Rightarrow l_{22} = a_{22} - l_{21}u_{12} = a_{22} - \mathrm{a}_{21}\left(\frac{a_{12}}{a_{11}}\right)$$

$$l_{31}u_{12} + l_{32} = a_{32} \Rightarrow l_{32} = a_{32} - l_{31}u_{12} = a_{32} - \mathrm{a}_{31}\left(\frac{a_{12}}{a_{11}}\right)$$

(d) Compare 2nd row

$$l_{21}\,u_{13}+l_{22}\,u_{23} = a_{23}$$

$$\Rightarrow l_{22}\,u_{23} = a_{23} - l_{21}\,u_{13}$$

$$\Rightarrow u_{23} = \frac{1}{l_{22}}\,(a_{23} - l_{21}\,u_{13})$$

(e) Compare 3rd column

$$l_{31}\,u_{13}+l_{32}\,u_{23}+l_{33} = a_{33}$$

$$\Rightarrow l_{33} = a_{33} - l_{31}\,u_{13} - l_{32}\,u_{23}$$

So on comparing all the elements of L and U are known and hence put the value **A (= LU)** in **AX = B**

(LU) X = B

L (UX) = B

Assume UX = Y, where Y = $\begin{bmatrix} y_1 \\ y_2 \\ y_3 \end{bmatrix}$

L Y = B

$$\begin{bmatrix} l_{11} & 0 & 0 \\ l_{21} & l_{22} & 0 \\ l_{31} & l_{32} & l_{33} \end{bmatrix} \begin{bmatrix} y_1 \\ y_2 \\ y_3 \end{bmatrix} = \begin{bmatrix} b_1 \\ b_2 \\ b_3 \end{bmatrix}$$

$$\begin{bmatrix} l_{11}\,y_1 \\ l_{21}\,y_1+l_{22}\,y_2 \\ l_{31}\,y_1+l_{32}\,y_2+l_{33}\,y_3 \end{bmatrix} = \begin{bmatrix} b_1 \\ b_2 \\ b_3 \end{bmatrix}$$

On comparing,

$$y_1 = \frac{l_{11}}{b_1}$$

$$y_2 = \frac{1}{l_{22}}(b_2 - l_{21}\,y_1)$$

$$y_3 = \frac{1}{l_{33}}(b_3 - l_{31}\,y_1 - l_{31}\,y_2)$$

$$\text{Or} \begin{bmatrix} y_1 \\ y_2 \\ y_3 \end{bmatrix} = \begin{bmatrix} \dfrac{l_{11}}{b_1} \\ \dfrac{1}{l_{22}}(b_2 - l_{21}\, y_1) \\ \dfrac{1}{l_{33}}(b_3 - l_{31}\, y_1 - l_{31}\, y_2) \end{bmatrix}$$

Put the value of Y in **U X = Y**

$$\begin{bmatrix} 1 & u_{12} & u_{13} \\ 0 & 1 & u_{23} \\ 0 & 0 & 1 \end{bmatrix} \begin{bmatrix} x_1 \\ x_2 \\ x_3 \end{bmatrix} = \begin{bmatrix} y_1 \\ y_2 \\ y_3 \end{bmatrix}$$

$$\begin{bmatrix} 1 & u_{12} & u_{13} \\ 0 & 1 & u_{23} \\ 0 & 0 & 1 \end{bmatrix} \begin{bmatrix} x_1 \\ x_2 \\ x_3 \end{bmatrix} = \begin{bmatrix} y_1 \\ y_2 \\ y_3 \end{bmatrix}$$

$$\begin{bmatrix} x_1 + u_{12}\, x_2 + u_{13}\, x_3 \\ x_2 + u_{23}\, x_3 \\ x_3 \end{bmatrix} = \begin{bmatrix} y_1 \\ y_2 \\ y_3 \end{bmatrix}$$

$$x_3 = y_3$$

$$x_2 = y_2 - u_{23}\, x_3$$

$$x_1 = y_1 - u_{12}\, x_2 - u_{13}\, x_3$$

Example 2.10: Through Crout's method, solve the following equations

3x + 2y + 7z = 4

2x + 3y + z = 5

3x + 4y + z = 7

Solution: The given system of equations

3x + 2y + 7z = 4

2x + 3y + z = 5

3x + 4y + z = 7

So write this system of equations in the form **AX = B** as

in which, $A = \begin{bmatrix} 3 & 2 & 7 \\ 2 & 3 & 1 \\ 3 & 4 & 1 \end{bmatrix}, X = \begin{bmatrix} x \\ y \\ z \end{bmatrix} and\ B = \begin{bmatrix} 4 \\ 5 \\ 7 \end{bmatrix}$

Let, **A = LU**

where L and U are the lower and upper triangular matrix respectively.

$$\begin{bmatrix} l_{11} & 0 & 0 \\ l_{21} & l_{22} & 0 \\ l_{31} & l_{32} & l_{33} \end{bmatrix} and\ U = \begin{bmatrix} 1 & u_{12} & u_{13} \\ 0 & 1 & u_{23} \\ 0 & 0 & 1 \end{bmatrix}$$

A = L U

$$\begin{bmatrix} 3 & 2 & 7 \\ 2 & 3 & 1 \\ 3 & 4 & 1 \end{bmatrix} = \begin{bmatrix} l_{11} & 0 & 0 \\ l_{21} & l_{22} & 0 \\ l_{31} & l_{32} & l_{33} \end{bmatrix} \begin{bmatrix} 1 & u_{12} & u_{13} \\ 0 & 1 & u_{23} \\ 0 & 0 & 1 \end{bmatrix}$$

Multiply the matrix on the right hand side,

$$\begin{bmatrix} 3 & 2 & 7 \\ 2 & 3 & 1 \\ 3 & 4 & 1 \end{bmatrix} = \begin{bmatrix} l_{11} & 0 & 0 \\ l_{21} & l_{22} & 0 \\ l_{31} & l_{32} & l_{33} \end{bmatrix} \begin{bmatrix} 1 & u_{12} & u_{13} \\ 0 & 1 & u_{23} \\ 0 & 0 & 1 \end{bmatrix}$$

Comparing and equating corresponding elements in the matrix of both sides as

(a) Compare 1st column

$$l_{11} = 3 l_{21} = 2 l_{31} = 3$$

(b) Compare 1st row and put above values

$$l_{11}\ u_{12} = a_{12} \Rightarrow u_{12} = \frac{a_{12}}{l_{11}} = \frac{2}{3}$$

$$l_{11}\ u_{13} = a_{13} \Rightarrow u_{13} = \frac{a_{13}}{l_{11}} = \frac{7}{3}$$

(c) Compare 2nd column

$$l_{21} u_{12} + l_{22} = a_{22} \Rightarrow l_{22} = a_{22} - l_{21} u_{12} = 3 - 2 \left(\frac{2}{3}\right) = \frac{5}{3}$$

$$l_{31}u_{12}+l_{32} = a_{32} \Rightarrow l_{32} = a_{32} - l_{31}u_{12} = a_{32} - a_{31}\left(\frac{a_{12}}{a_{11}}\right) = 4 - 3\left(\frac{2}{3}\right) = 2$$

(d) Compare 2nd row

$$l_{21}u_{13}+l_{22}u_{23} = a_{23}$$
$$\Rightarrow l_{22}u_{23} = a_{23} - l_{21}u_{13}$$
$$\Rightarrow u_{23} = \frac{1}{l_{22}}(a_{23} - l_{21}u_{13})$$
$$\Rightarrow u_{23} = \frac{1}{5/3}\left(1 - 2\left(7/3\right)\right) = \frac{-11}{5}$$

(e) Compare 3rd column

$$l_{31}u_{13}+l_{32}u_{23}+l_{33} = a_{33}$$
$$\Rightarrow l_{33} = a_{33} - l_{31}u_{13} - l_{32}u_{23}$$
$$\Rightarrow l_{33} = 1-3\left(\frac{7}{3}\right)-2\left(\frac{-11}{5}\right) = \frac{-8}{5}$$

So on comparing all the elements of L and U are known and hence put the value **A (= LU)** in **AX = B**

(LU) X = B

L (UX) = B

Assume UX = Y, where Y = $\begin{bmatrix} y_1 \\ y_2 \\ y_3 \end{bmatrix}$

L Y = B

$$\begin{bmatrix} l_{11} & 0 & 0 \\ l_{21} & l_{22} & 0 \\ l_{31} & l_{32} & l_{33} \end{bmatrix}\begin{bmatrix} y_1 \\ y_2 \\ y_3 \end{bmatrix} = \begin{bmatrix} b_1 \\ b_2 \\ b_3 \end{bmatrix}$$

$$\begin{bmatrix} 3 & 0 & 0 \\ 2 & 5/3 & 0 \\ 3 & 2 & -8/3 \end{bmatrix} \begin{bmatrix} y_1 \\ y_2 \\ y_3 \end{bmatrix} = \begin{bmatrix} 4 \\ 5 \\ 7 \end{bmatrix}$$

$$\begin{bmatrix} l_{11}\, y_1 \\ l_{21}\, y_1 + l_{22}\, y_2 \\ l_{31}\, y_1 + l_{32}\, y_2 + l_{33}\, y_3 \end{bmatrix} = \begin{bmatrix} b_1 \\ b_2 \\ b_3 \end{bmatrix}$$

On comparing,

$$y_1 = \frac{l_{11}}{b_1} = \frac{4}{3}$$

$$y_2 = \frac{1}{l_{22}}(b_2 - l_{21}\, y_1) = \frac{1}{5/3}\left(5 - 2\left(4/3\right)\right) = \frac{7}{5}$$

$$y_3 = \frac{1}{l_{33}}(b_3 - l_{31}\, y_1 - l_{31}\, y_2) = \frac{1}{8/5}\left(7 - 3\left(4/3\right) - 3\left(7/5\right)\right) = \frac{-1}{8}$$

Or = $\begin{bmatrix} y_1 \\ y_2 \\ y_3 \end{bmatrix} = \begin{bmatrix} 4/3 \\ 7/5 \\ -1/8 \end{bmatrix}$

Put the value of Y in **U X = Y**

$$\begin{bmatrix} 1 & u_{12} & u_{13} \\ 0 & 1 & u_{23} \\ 0 & 0 & 1 \end{bmatrix} \begin{bmatrix} x_1 \\ x_2 \\ x_3 \end{bmatrix} = \begin{bmatrix} y_1 \\ y_2 \\ y_3 \end{bmatrix}$$

$$\begin{bmatrix} 1 & 2/3 & 7/3 \\ 0 & 1 & -11/5 \\ 0 & 0 & 1 \end{bmatrix} \begin{bmatrix} x \\ y \\ z \end{bmatrix} = \begin{bmatrix} 4/3 \\ 7/5 \\ -1/8 \end{bmatrix}$$

$$\begin{bmatrix} x+\frac{2}{3}y+\frac{7}{3}z \\ y-\frac{11}{5}z \\ z \end{bmatrix} = \begin{bmatrix} 4/3 \\ 7/5 \\ -1/8 \end{bmatrix}$$

$$z=\frac{4}{3}$$

$$y-\frac{11}{5}z=\frac{7}{5} \Rightarrow y=\frac{9}{8}$$

$$x+\frac{2}{3}y+\frac{7}{3}z=\frac{-1}{8} \Rightarrow x=\frac{7}{8}$$

Hence the solution is as

$$\begin{bmatrix} x \\ y \\ z \end{bmatrix} = \begin{bmatrix} \frac{7}{8} \\ \frac{9}{8} \\ \frac{4}{3} \end{bmatrix}$$

2.9 Cholesky's Triangularisation Method

This method is also called '**square root method**'.

Suppose in case of real matrix A= $\begin{bmatrix} a_{11} & a_{12} & a_{13} \\ a_{21} & a_{22} & a_{23} \\ a_{31} & a_{32} & a_{33} \end{bmatrix}$

A = L L'

where L = lower triangular matrix = $\begin{bmatrix} l_{11} & 0 & 0 \\ l_{21} & l_{22} & 0 \\ l_{31} & l_{32} & l_{33} \end{bmatrix}$ and

$$L' = \text{transpose of } L = \begin{bmatrix} l_{11} & l_{21} & l_{31} \\ 0 & l_{22} & l_{32} \\ 0 & 0 & l_{33} \end{bmatrix}$$

Take inverse of above equation, yields

$A^{-1} = (L\,L')^{-1} = L^{-1}\,(L')^{-1} = L^{-1}\,(L^{-1})'$

$$\begin{bmatrix} a_{11} & a_{12} & a_{13} \\ a_{21} & a_{22} & a_{23} \\ a_{31} & a_{32} & a_{33} \end{bmatrix} = \begin{bmatrix} l_{11} & 0 & 0 \\ l_{21} & l_{22} & 0 \\ l_{31} & l_{32} & l_{33} \end{bmatrix} \begin{bmatrix} l_{11} & l_{21} & l_{31} \\ 0 & l_{22} & l_{32} \\ 0 & 0 & l_{33} \end{bmatrix}$$

$$\begin{bmatrix} a_{11} & a_{12} & a_{13} \\ a_{21} & a_{22} & a_{23} \\ a_{31} & a_{32} & a_{33} \end{bmatrix} = \begin{bmatrix} l_{11}^2 & l_{11}\,l_{21} & l_{11}\,l_{31} \\ l_{21}\,l_{11} & l_{21}^2 + l_{22}^2 & l_{21}\,l_{31} + l_{22}\,l_{32} \\ l_{31}\,l_{11} & l_{31}\,l_{21} + l_{32}\,l_{22} & l_{31}^2 + l_{32}^2 + l_{33}^2 \end{bmatrix}$$

By comparing and equating the matrix as

$$l_{11}^2 = a_{11} \Rightarrow l_{11} = \sqrt{a_{11}}$$

$$l_{11}\,l_{21} = a_{12} \Rightarrow l_{21} = \frac{a_{12}}{l_{11}}$$

$$l_{11}\,l_{31} = a_{13} \Rightarrow l_{31} = \frac{a_{13}}{l_{11}}$$

$$l_{21}^2 + l_{22}^2 = a_{22} \Rightarrow l_{22} = \sqrt{a_{22} - l_{21}^2}$$

$$l_{21}\,l_{31} + l_{22}\,l_{32} = a_{23} \Rightarrow l_{32} = \frac{1}{l_{22}}\left(a_{23} - l_{21}\,l_{31}\right)$$

$$l_{31}^2 + l_{32}^2 + l_{33}^2 = a_{33} \Rightarrow l_{22} = \sqrt{a_{33} - l_{31}^2 - l_{32}^2}$$

$$L = \begin{bmatrix} l_{11} & 0 & 0 \\ l_{21} & l_{22} & 0 \\ l_{31} & l_{32} & l_{33} \end{bmatrix} = \begin{bmatrix} \sqrt{a_{11}} & 0 & 0 \\ \dfrac{a_{12}}{l_{11}} & \sqrt{a_{22} - l_{21}^2} & 0 \\ \dfrac{a_{13}}{l_{11}} & \dfrac{1}{l_{22}}\left(a_{23} - l_{21}\,l_{31}\right) & \sqrt{a_{33} - l_{31}^2 - l_{32}^2} \end{bmatrix}$$

Example 2.11: Using Choleski's method, solve the system of equations i.e.,

x + 2y + 6z = 5

2x + 5y + 15z = 12

6x + 15y + 46z = 37

Solution:

A X = B (1)

$$\text{Matrix A}=\begin{bmatrix}1 & 2 & 6\\2 & 5 & 15\\6 & 15 & 46\end{bmatrix},\ X=\begin{bmatrix}x\\y\\z\end{bmatrix},\ B=\begin{bmatrix}5\\12\\37\end{bmatrix}$$

A = L L' (2)

where L (lower triangular matrix) = $\begin{bmatrix}l_{11} & 0 & 0\\l_{21} & l_{22} & 0\\l_{31} & l_{32} & l_{33}\end{bmatrix}$ and L' (transpose of L)

$$=\begin{bmatrix}l_{11} & l_{21} & l_{31}\\0 & l_{22} & l_{32}\\0 & 0 & l_{33}\end{bmatrix}$$

$$\begin{bmatrix}1 & 2 & 6\\2 & 5 & 15\\6 & 15 & 46\end{bmatrix}=\begin{bmatrix}l_{11} & 0 & 0\\l_{21} & l_{22} & 0\\l_{31} & l_{32} & l_{33}\end{bmatrix}\begin{bmatrix}l_{11} & l_{21} & l_{31}\\0 & l_{22} & l_{32}\\0 & 0 & l_{33}\end{bmatrix}$$

$$\begin{bmatrix}1 & 2 & 6\\2 & 5 & 15\\6 & 15 & 46\end{bmatrix}=\begin{bmatrix}l_{11}^2 & l_{11}l_{21} & l_{11}l_{31}\\l_{21}l_{11} & l_{21}^2+l_{22}^2 & l_{21}l_{31}+l_{22}l_{32}\\l_{31}l_{11} & l_{31}l_{21}+l_{32}l_{22} & l_{31}^2+l_{32}^2+l_{33}^2\end{bmatrix}$$

By comparing and equating the matrix as

$$l_{11}^2=a_{12} \Rightarrow l_{11}=\sqrt{a_{12}}=\sqrt{1}=1$$

$$l_{11}l_{21}=a_{12} \Rightarrow l_{21}=\frac{a_{12}}{l_{11}}=\frac{2}{1}=2$$

$$l_{11}l_{31}=a_{13} \Rightarrow l_{31}=\frac{a_{13}}{l_{11}}=\frac{6}{1}=6$$

$$l_{21}^{\ 2} + l_{22}^{\ 2} = a_{22} \Rightarrow l_{22} = \sqrt{a_{22} - l_{21}^{\ 2}} = \sqrt{5 - 2^2} = 1$$

$$l_{21}\, l_{31} + l_{22} l_{32} = a_{23} \Rightarrow l_{32} = \frac{1}{l_{22}}\left(a_{23} - l_{21}\, l_{31}\right) = \frac{1}{1}\left(15 - 2(6)\right) = 3$$

$$l_{31}^{\ 2} + l_{32}^{\ 2} + l_{33}^{\ 2} = a_{33} \Rightarrow l_{22} = \sqrt{a_{33} - l_{31}^{\ 2} - l_{32}^{\ 2}} = \sqrt{46 - 6^2 - 3^2} = 1$$

$$\mathrm{L} = \begin{bmatrix} l_{11} & 0 & 0 \\ l_{21} & l_{22} & 0 \\ l_{31} & l_{32} & l_{33} \end{bmatrix} = \begin{bmatrix} 1 & 0 & 0 \\ 2 & 1 & 0 \\ 6 & 3 & 1 \end{bmatrix}$$

$$\mathrm{L}' = \begin{bmatrix} l_{11} & l_{21} & l_{31} \\ 0 & l_{22} & l_{32} \\ 0 & 0 & l_{33} \end{bmatrix} = \begin{bmatrix} 1 & 2 & 6 \\ 0 & 1 & 3 \\ 0 & 0 & 1 \end{bmatrix}$$

Put the value **A (= LL')** in **AX = B**

(LL') X = B

L (L'X) = B

Assume L'X = Y (3)

$$\text{where } \mathrm{Y} = \begin{bmatrix} y_1 \\ y_2 \\ y_3 \end{bmatrix}$$

L Y = B

$$\begin{bmatrix} 1 & 0 & 0 \\ 2 & 1 & 0 \\ 6 & 3 & 1 \end{bmatrix} \begin{bmatrix} y_1 \\ y_2 \\ y_3 \end{bmatrix} = \begin{bmatrix} 5 \\ 12 \\ 37 \end{bmatrix}$$

Left hand side multiplication yields,

$$\begin{bmatrix} y_1 \\ 2\,y_1 + 1\,y_2 \\ 6\,y_1 + 3\,y_2 + 1\,y_3 \end{bmatrix} = \begin{bmatrix} 5 \\ 12 \\ 37 \end{bmatrix}$$

$$y_1 = 5$$

On comparing, $2y_1 + 1y_2 = 12 \Rightarrow y_2 = 2$

$$6y_1 + 3y_2 + 1y_3 = 37 \Rightarrow y_3 = 1$$

$$\begin{bmatrix} y_1 \\ y_2 \\ y_3 \end{bmatrix} = \begin{bmatrix} 5 \\ 2 \\ 1 \end{bmatrix}$$

From equation 3,
L' X = Y

$$\begin{bmatrix} 1 & 2 & 6 \\ 0 & 1 & 3 \\ 0 & 0 & 1 \end{bmatrix} \begin{bmatrix} x \\ y \\ z \end{bmatrix} = \begin{bmatrix} 5 \\ 2 \\ 1 \end{bmatrix}$$

$$x + 2y + 6z = 5$$
$$y + 3z = 2$$
$$z = 1$$

Put the value of z to find y and then x. The solution is

$$\begin{bmatrix} x \\ y \\ z \end{bmatrix} = \begin{bmatrix} 1 \\ -1 \\ 1 \end{bmatrix}$$

2.10 Thomas Algorithm for Tridiagonal System

A **band matrix** is a matrix in which non-zero elements are confined to a diagonal band, which comprises the main diagonal and zero or more diagonals on either side. For example,

Banded matrix with non-zero diagonal elements; $\begin{bmatrix} 1 & 0 & 0 \\ 0 & 2 & 0 \\ 0 & 0 & 3 \end{bmatrix}$, in which main diagonal elements are 1, 2 and 3.

Banded matrix with non-zero diagonal elements and more diagonals on side, $\begin{bmatrix} 1 & 4 & 0 \\ 6 & 2 & 5 \\ 0 & 7 & 3 \end{bmatrix}$. In this matrix, the elements 4 and 5 is **super-diagonal** (above main diagonal) and the number of super-diagonals is called **upper bandwidth**, and in this example the upper bandwidth is 1, comprising elements 4 and 5. The elements 6 and 7 is **sub-diagonal** (below main diagonal) and the number of sub-diagonals is called **lower bandwidth**, and in this example the lower bandwidth is 1, comprising elements 6 and 7.

In **symmetric matrix**, the super-diagonals and sub-diagonals are equal. For example, in the matrix $\begin{bmatrix} 1 & 2 & 3 & 4 \\ 5 & 1 & 2 & 3 \\ 6 & 5 & 1 & 2 \\ 7 & 6 & 5 & 1 \end{bmatrix}$, where element 1 represents main diagonal; elements 2 and 3 represents 1st and 2nd upper-diagonals respectively and upper bandwidth is 2; elements 5 and 6 represents 1st and 2nd lower-diagonals respectively and lower bandwidth is 2; hence this matrix is symmetric having total of 5 diagonals.

A **tridiagonal matrix** is a band matrix which has non-zero elements only on the main diagonal, 1st upper-diagonal (above main diagonal) and 1st lower-diagonal (below main diagonal). For example, consider matrix of order 4X4,

$$\begin{bmatrix} a_1 & b_1 & 0 & 0 \\ c_1 & a_2 & b_2 & 0 \\ 0 & c_2 & a_3 & b_3 \\ 0 & 0 & c_3 & a_4 \end{bmatrix} \text{ such as } \begin{bmatrix} 1 & 5 & 0 & 0 \\ 8 & 2 & 6 & 0 \\ 0 & 9 & 3 & 7 \\ 0 & 0 & 10 & 4 \end{bmatrix}$$

Tridiagonal matrix algorithm (TDMA) is also called **Thomas' algorithm**, is the result of applying Gauss elimination to the tridiagonal system of equations i.e., system of equations in which non-zero coefficients are present on the main diagonal, 1st upper-diagonal and 1st lower diagonal and the corresponding the system of equations is called **tridiagonal system of equations**.

Consider a tridiagonal system with n equations with unknowns x_1, x_2, ... x_{n-1}, x_n as mentioned below.

$$\begin{bmatrix} a_1 & b_1 & & & \\ c_2 & a_2 & b_2 & & \\ & c_3 & \ddots & \ddots & \\ & & \ddots & a_{n-1} & b_{n-1} \\ & & & c_n & a_n \end{bmatrix} \begin{bmatrix} x_1 \\ x_2 \\ \vdots \\ x_{n-1} \\ x_n \end{bmatrix} = \begin{bmatrix} d_1 \\ d_2 \\ \vdots \\ d_{n-1} \\ d_n \end{bmatrix}$$

From the matrix, it must be noted that $c_2 = 0$ and $b_n = 0$. Thus the matrix is

$c_i x_{i-1} + \alpha_i x_i + b_i x_{i+1} = d_i$

The ith unknown can be written as function of (i+1)[th] unknown, i.e.,

$x_i = \lambda_i x_{i+1} + \mu_i$

$x_{i-1} = \lambda_{i-1} x_i + \mu_{i-1}$

where λ and μ are the coefficients.

Advantages

(1) Numerical computation is very easy, even for large number of system of equations, for stable condition i.e., $|a_i|>|b_i|+|c_i|$ for each value of i.

(2) This type of system is most common and most convert the general system of equations into this system of equations.

Disadvantages

This type of system fails when the equation becomes $a_i - b_i \lambda_{i-1} = 0$ and for singular trigonal matrix $\begin{bmatrix} 0 & 1 & 1 \\ 1 & 0 & 1 \\ 1 & 1 & 0 \end{bmatrix}$ where determinant is zero & no inverse and in some cases of non-singular also i.e., $\begin{bmatrix} 4 & 1 & 1 \\ -7 & -5 & 0 \\ 1 & 1 & -1 \end{bmatrix}$ where determinant is non-zero & has inverse. So to solve the Jacobi system of equation pivoting has to be done.

Example 2.12: Solve the given system of equations through Thomas' algorithm

$3x_1 - x_2 + 0\,x_3 = -1$

$-x_1 + 3x_2 - x_3 = 7$

$0\,x_1 - x_2 + 3x_3 = 7$

Solution: The given system of equations

$3x_1 - x_2 + 0\, x_3 = -1$ (1)

$-x_1 + 3x_2 - x_3 = 7$ (2)

$0\, x_1 - x_2 + 3x_3 = 7$ (3)

The corresponding matrix form is

$$\begin{bmatrix} a_1 & b_1 & 0 \\ c_2 & a_2 & b_1 \\ 0 & c_3 & a_3 \end{bmatrix}\begin{bmatrix} x_1 \\ x_2 \\ x_3 \end{bmatrix} = \begin{bmatrix} d_1 \\ d_2 \\ d_3 \end{bmatrix}$$

$$\begin{bmatrix} 3 & -1 & 0 \\ -1 & 3 & -1 \\ 0 & -1 & 3 \end{bmatrix}\begin{bmatrix} x_1 \\ x_2 \\ x_3 \end{bmatrix} = \begin{bmatrix} -1 \\ 7 \\ 7 \end{bmatrix}$$

Divide the equation (1) by 3 and the equation becomes (1.1) as

$$x_1 - \frac{1}{3}x_2 = -\frac{1}{3}$$

or

$x_1 - \lambda_1\, x_2 = \mu_1$ (1.1)

where $\lambda_1 = -\dfrac{1}{3} = -0.33$ and $\mu_1 = -\dfrac{1}{3} = -0.33$

The new system of equations

$x_1 - 0.33\, x_2 = -\, 0.33$ (1.1)

$-x_1 + 3x_2 - x_3 = 7$ (2)

$0\, x_1 - x_2 + 3x_3 = 7$ (3)

The matrix is

$$\begin{bmatrix} 1 & -0.33 & 0 \\ -1 & 3 & -1 \\ 0 & -1 & 3 \end{bmatrix}\begin{bmatrix} x_1 \\ x_2 \\ x_3 \end{bmatrix} = \begin{bmatrix} -0.33 \\ 7 \\ 7 \end{bmatrix}$$

For equation (2), convert MX=D into UX=P

Operate [equation (2) – (–1) equation (1.1)] in order to eliminate x_1 from equation (2) yields,

$x_2 + \lambda_2 x_3 = \mu_2$

where $\lambda_2 = 0.375$ and $\mu_2 = 2.5$

$x_2 + 0.375\, x_3 = 2.5$ (2.1)

The new system of equations

$x_1 - 0.33\, x_2 = -\, 0.33$ (1.1)

$x_2 + 0.375\, x_3 = 2.5$ (2.1)

$0\, x_1 - x_2 + 3x_3 = 7$ (3)

The matrix is

$$\begin{bmatrix} 1 & -0.33 & 0 \\ 0 & 1 & 0.375 \\ 0 & -1 & 3 \end{bmatrix} \begin{bmatrix} x_1 \\ x_2 \\ x_3 \end{bmatrix} = \begin{bmatrix} -0.33 \\ 2.5 \\ 7 \end{bmatrix}$$

For equation (3), convert MX=D into UX=P

Operate [equation (3) – (–1) equation (1.1)] in order to eliminate x_2 from equation (3) yields,

$x_3 = \mu_3$

where $\mu_3 = 3.619$

$x_3 = 3.619$ (3.1)

The new system of equations

$x_1 - 0.33\, x_2 = -\, 0.33$ (1.1)

$x_2 + 0.375\, x_3 = 2.5$ (2.1)

$x_3 = 3.619$ (3.1)

The matrix is

$$\begin{bmatrix} 1 & -0.33 & 0 \\ 0 & 1 & 0.375 \\ 0 & 0 & 1 \end{bmatrix} \begin{bmatrix} x_1 \\ x_2 \\ x_3 \end{bmatrix} = \begin{bmatrix} -0.33 \\ 2.5 \\ 3.619 \end{bmatrix}$$

Upon back substitution of x_3 in equation (2.1) yields, $x_2 = 3.857$

And further upon back substitution of x_2 and x_3 in equation (1.1) yields, $x_1 = 0.952$

Hence the solution is $x_1 = 0.952$, $x_2 = 3.857$, $x_3 = 3.619$

2.11 Jacobi's Iteration Method

It is iterative method in order to find the solution of system of linear equations. This method is also called '**method of simultaneous displacements**'. Consider the

$$\begin{aligned} a_{11}x_1 + a_{12}x_2 + a_{13}x_3 + \cdots + a_{1n}x_n &= b_1 \\ a_{11}x_1 + a_{12}x_2 + a_{13}x_3 + \cdots + a_{1n}x_n &= b_2 \\ a_{11}x_1 + a_{12}x_2 + a_{13}x_3 + \cdots + a_{1n}x_n &= b_3 \\ \vdots \quad\quad \vdots \quad\quad \vdots \quad\quad\quad \vdots \;\; \vdots \quad &\quad \vdots \\ a_{n1}x_1 + a_{n2}x_2 + a_{n3}x_3 + \cdots + a_{nn}x_n &= b_n \end{aligned}$$

Assumption, $a_{ii} \neq 0$ for $i = 1, 2, 3, \ldots, n$. If this condition does not satisfied, then need to re-arrange the equations to fulfill the criteria.

The above equations can be re-written as

$$\begin{aligned} x_1 &= \frac{b_1}{a_{11}} - \frac{a_{12}}{a_{11}}x_2 - \frac{a_{13}}{a_{11}}x_3 - \cdots - \frac{a_{1n}}{a_{11}}x_n \\ x_2 &= \frac{b_2}{a_{22}} - \frac{a_{21}}{a_{22}}x_1 - \frac{a_{23}}{a_{22}}x_3 - \cdots - \frac{a_{2n}}{a_{22}}x_n \\ x_3 &= \frac{b_3}{a_{33}} - \frac{a_{31}}{a_{33}}x_1 - \frac{a_{32}}{a_{33}}x_2 - \cdots - \frac{a_{3n}}{a_{33}}x_n \\ \vdots \quad & \quad \vdots \quad\quad \vdots \quad\quad\quad \vdots \quad\quad\quad \vdots \quad \vdots \\ x_n &= \frac{b_n}{a_{nn}} - \frac{a_{n1}}{a_{nn}}x_1 - \frac{a_{n2}}{a_{nn}}x_2 - \cdots - \frac{a_{n,n-1}}{a_{nn}}x_{n-1} \end{aligned}$$

Start a solution with an approximation i.e., $x_1^{(0)}, x_2^{(0)}, x_3^{(0)}, \ldots\ldots\ldots x_n^{(0)}$ are the initial approximate values of unknown variables $x_1, x_2, x_3, \ldots\ldots\ldots x_n$

where $x_1^{(0)}$ represents $x_1 = 0$ and so on.

Note: It is not necessary to take the initial values as zero, it may be reasonable guess.

Substitute the initial values in the above set of system equations and it becomes 1st approximation as

$$x_1^{(1)} = \frac{b_1}{a_{11}} - \frac{a_{12}}{a_{11}} x_2^{(0)} - \frac{a_{13}}{a_{11}} x_3^{(0)} - \cdots - \frac{a_{1n}}{a_{11}} x_n^{(0)}$$

$$x_2^{(1)} = \frac{b_2}{a_{22}} - \frac{a_{21}}{a_{22}} x_1^{(0)} - \frac{a_{23}}{a_{22}} x_3^{(0)} - \cdots - \frac{a_{2n}}{a_{22}} x_n^{(0)}$$

$$x_3^{(1)} = \frac{b_3}{a_{33}} - \frac{a_{31}}{a_{33}} x_1^{(0)} - \frac{a_{32}}{a_{33}} x_2^{(0)} - \cdots - \frac{a_{3n}}{a_{33}} x_n^{(0)}$$

$$\vdots \quad \vdots \quad \vdots \quad \vdots \quad \vdots \quad \vdots$$

$$x_n^{(1)} = \frac{b_n}{a_{nn}} - \frac{a_{n1}}{a_{nn}} x_1^{(0)} - \frac{a_{n2}}{a_{nn}} x_2^{(0)} - \cdots - \frac{a_{n,n-1}}{a_{nn}} x_{n-1}^{(0)}$$

Then 2nd iteration / approximation; $x_1^{(1)}, x_2^{(1)}, x_3^{(1)}, \ldots\ldots\ldots x_n^{(1)}$ are the next values of unknown variables $x_1, x_2, x_3, \ldots\ldots\ldots x_n$

Substitute these 1st iteration values in the above set of system equations and it becomes 2nd approximation as

$$x_1^{(2)} = \frac{b_1}{a_{11}} - \frac{a_{12}}{a_{11}} x_2^{(1)} - \frac{a_{13}}{a_{11}} x_3^{(1)} - \cdots - \frac{a_{1n}}{a_{11}} x_n^{(1)}$$

$$x_2^{(2)} = \frac{b_2}{a_{22}} - \frac{a_{21}}{a_{22}} x_1^{(1)} - \frac{a_{23}}{a_{22}} x_3^{(1)} - \cdots - \frac{a_{2n}}{a_{22}} x_n^{(1)}$$

$$x_3^{(2)} = \frac{b_3}{a_{33}} - \frac{a_{31}}{a_{33}} x_1^{(1)} - \frac{a_{32}}{a_{33}} x_2^{(1)} - \cdots - \frac{a_{3n}}{a_{33}} x_n^{(1)}$$

$$\vdots \quad \vdots \quad \vdots \quad \vdots \quad \vdots \quad \vdots$$

$$x_1^{(2)} = \frac{b_n}{a_{nn}} - \frac{a_{n1}}{a_{nn}} x_1^{(1)} - \frac{a_{n2}}{a_{nn}} x_2^{(1)} - \cdots - \frac{a_{n,n-1}}{a_{nn}} x_{n-1}^{(1)}$$

Then proceed to 3rd & 4th approximation and so on in order to get the exact solution. It must ensure that the difference between the two successive values of unknown variables should be approximately same.

Example 2.13: Solve the system of equations x + 2y + z = 12; x + 4y+ 2z = 15; x + 2y + 5z = 20 through Jacobi's iteration method.

Solution: The given system of equations is

$5x + 2y + z = 12$ (1)

$x + 4y + 2z = 15$ (2)

$x + 2y + 5z = 20$ (3)

Re-write the equations 1, 2 & 3 as

$$x = \frac{1}{5}(12 - 2y - z)$$

$$y = \frac{1}{4}(15 - x - 2z)$$

$$z = \frac{1}{5}(20 - x - 2y)$$

Start a solution with an approximation i.e., $x^{(0)} = 0$, $y^{(0)} = 0$ and $z^{(0)} = 0$ are the initial approximate values of unknown variables x, y and z.

Note: $x^{(0)}$ can also be written as x_0. For this it must ensure that unknown variables are given different letter.

1st Iteration:

$$x^{(1)} = \frac{1}{5}(12 - 2y^{(0)} - z^{(0)}) = \frac{1}{5}(12 - 2(0) - 0) = 2.4$$

$$y^{(1)} = \frac{1}{4}(15 - x^{(0)} - 2z^{(0)}) = \frac{1}{4}(15 - 0 - 2(0)) = 3.75$$

$$z^{(1)} = \frac{1}{5}(20 - x^{(0)} - 2y^{(0)}) = \frac{1}{5}(20 - 0 - 2(0)) = 4$$

2nd Iteration:

$$x^{(2)} = \frac{1}{5}(12 - 2y^{(1)} - z^{(1)}) = \frac{1}{5}(12 - 2(3.75) - 4) = 0.1$$

$$y^{(2)} = \frac{1}{4}(15 - x^{(1)} - 2z^{(1)}) = \frac{1}{4}(15 - 2.4 - 2(4)) = 1.15$$

$$z^{(2)} = \frac{1}{5}(20 - x^{(1)} - 2y^{(1)}) = \frac{1}{5}(20 - 2.4 - 2\,(3.75)) = 2.02$$

3rd Iteration:

$$x^{(3)} = \frac{1}{5}(12 - 2y^{(2)} - z^{(2)}) = \frac{1}{5}(12 - 2\,(1.15) - 2.02) = 1.536$$

$$y^{(3)} = \frac{1}{4}(15 - x^{(2)} - 2z^{(2)}) = \frac{1}{4}(15 - 0.1 - 2\,(2.02)) = 2.715$$

$$z^{(3)} = \frac{1}{5}(20 - x^{(2)} - 2y^{(2)}) = \frac{1}{5}(20 - 0.1 - 2\,(1.15)) = 3.52$$

4th Iteration:

$$x^{(4)} = \frac{1}{5}(12 - 2y^{(3)} - z^{(3)}) = \frac{1}{5}(12 - 2\,(2.715) - 3.52) = 0.61$$

$$y^{(4)} = \frac{1}{4}(15 - x^{(3)} - 2z^{(3)}) = \frac{1}{4}(15 - 1.536 - 2\,(3.52)) = 1.606$$

$$z^{(4)} = \frac{1}{5}(20 - x^{(3)} - 2y^{(3)}) = \frac{1}{5}(20 - 1.536 - 2\,(2.715)) = 2.607$$

5th Iteration:

$$x^{(5)} = \frac{1}{5}(12 - 2y^{(4)} - z^{(4)}) = \frac{1}{5}(12 - 2\,(1.606) - 2.607) = 1.236$$

$$y^{(5)} = \frac{1}{4}(15 - x^{(4)} - 2z^{(4)}) = \frac{1}{4}(15 - 0.61 - 2\,(2.607)) = 2.294$$

$$z^{(5)} = \frac{1}{5}(20 - x^{(4)} - 2y^{(4)}) = \frac{1}{5}(20 - 0.61 - 2\,(1.606)) = 3.236$$

6th Iteration:

$$x^{(6)} = \frac{1}{5}(12 - 2y^{(5)} - z^{(5)}) = \frac{1}{5}(12 - 2\,(2.294) - 3.236) = 0.835$$

$$y^{(6)} = \frac{1}{4}(15 - x^{(5)} - 2z^{(5)}) = \frac{1}{4}(15 - 1.236 - 2\,(3.236)) = 1.823$$

$$z^{(6)} = \frac{1}{5}(20 - x^{(5)} - 2y^{(5)}) = \frac{1}{5}(20 - 1.236 - 2\,(2.294)) = 2.835$$

7th Iteration:

$$x^{(7)} = \frac{1}{5}(12 - 2y^{(6)} - z^{(6)}) = \frac{1}{5}(12 - 2\,(1.823) - 2.835) = 1.104$$

$$y^{(7)} = \frac{1}{4}(15 - x^{(6)} - 2z^{(6)}) = \frac{1}{4}(15 - 0.835 - 2\,(2.835)) = 2.124$$

$$z^{(7)} = \frac{1}{5}(20 - x^{(6)} - 2y^{(6)}) = \frac{1}{5}(20 - 0.835 - 2\,(1.823) = 3.104$$

Proceeding with the same way,

8th iteration: $x^{(8)} = 0.930$, $y^{(8)} = 1.922$, $z^{(8)} = 2.930$

9th iteration: $x^{(9)} = 1.045$, $y^{(9)} = 2.052$, $z^{(9)} = 3.045$

10th iteration: $x^{(10)} = 0.970$, $y^{(10)} = 1.966$, $z^{(10)} = 2.970$

11th iteration: $x^{(11)} = 1.020$, $y^{(11)} = 2.022$, $z^{(11)} = 3.020$

12th iteration: $x^{(12)} = 0.987$, $y^{(12)} = 1.985$, $z^{(12)} = 2.987$

13th iteration: $x^{(13)} = 1.009$, $y^{(13)} = 2.010$, $z^{(13)} = 3.009$

14th iteration: $x^{(14)} = 0.994$, $y^{(14)} = 1.993$, $z^{(14)} = 2.994$

From iteration 13th and 14th, it is clear that the difference between these iterations is approximately same, so need to stop the iteration. Hence the solution of system of equations is

$x = 0.994$, $y = 1.993$, $z = 2.994$

Example 2.14: Solve the system of equations $10a - 2b - c - d = 3$; $-2a + 10b - c - d = 15$; $-a - b + 10c - 2d = 27$; $-a - b - 2c + 10d = -9$, using Jacobi's iteration method.

Solution: The given system of equations is

$$10a - 2b - c - d = 3 \quad (1)$$

$$-2a + 10b - c - d = 15 \quad (2)$$

$$-a - b + 10c - 2d = 27 \quad (3)$$

$$-a - b - 2c + 10d = -9 \quad (4)$$

Re-write the equations 1, 2, 3 & 4 as

$$a = \frac{1}{10}(3 + 2b + c + d)$$

$$b = \frac{1}{10}(15 + 2a + c + d)$$

$$c = \frac{1}{10}(27 + a + b + 2d)$$

$$d = \frac{1}{10}(-9 + a + b + 2c)$$

Start a solution with an approximation i.e., $a_0 = 0$ (or $a^{(0)} = 0$), $b_0 = 0$, $c_0 = 0$, $d_0 = 0$, are the initial approximate values of unknown variables a, b, c and d.

1st iteration

$$a_1 = \frac{1}{10}(3 + 2b_0 + c_0 + d_0) = \frac{1}{10}(3 + 2\,(0) + 0 + 0) = 0.3$$

$$b_1 = \frac{1}{10}(15 + 2a_0 + c_0 + d_0) = \frac{1}{10}(15 + 2\,(0) + 0 + 0) = 1.5$$

$$c_1 = \frac{1}{10}(27 + a_0 + b_0 + 2d_0) = \frac{1}{10}(27 + 0 + 0 + 2\,(0)) = 2.7$$

$$d_1 = \frac{1}{10}(-9 + a_0 + b_0 + 2c_0) = \frac{1}{10}(-9 + 0 + 0 + 2\,(0)) = -0.9$$

2nd iteration

$$a_2 = \frac{1}{10}(3 + 2b_1 + c_1 + d_1) = \frac{1}{10}(3 + 2\,(1.5) + 2.7 + (-0.9)) = 0.78$$

$$b_2 = \frac{1}{10}(15 + 2a_1 + c_1 + d_1) = \frac{1}{10}(15 + 2\,(0.3) + 2.7 + (-0.9)) = 1.74$$

$$c_2 = \frac{1}{10}(27 + a_1 + b_1 + 2d_1) = \frac{1}{10}(27 + 0.3 + 1.5 + 2\,(-0.9)) = 2.7$$

$$d_2 = \frac{1}{10}(-9 + a_1 + b_1 + 2c_1) = \frac{1}{10}(-9 + 0.3 + 1.5 + 2\,(2.7)) = -0.18$$

3rd iteration

$$a_3 = \frac{1}{10}(3 + 2b_2 + c_2 + d_2) = \frac{1}{10}(3 + 2\,(1.74) + 2.7 + (-0.18)) = 0.9$$

$$b_3 = \frac{1}{10}\,(15 + 2a_2 + c_2 + d_2) = \frac{1}{10}(15 + 2\,(0.78) + 2.7 + (-0.18)) = 1.908$$

$$c_3 = \frac{1}{10}\,(27 + a_2 + b_2 + 2d_2) = \frac{1}{10}(27 + 0.78 + 1.74 + 2\,(-0.18)) = 2.916$$

$$d_3 = \frac{1}{10}(-9 + a_2 + b_2 + 2c_2) = \frac{1}{10}\,(-9 + 0.78 + 1.74 + 2\,(2.7)) = -0.108$$

Proceeding with the same way,

4th iteration: $a_4 = 0.9624$, $b_4 = 1.9608$, $c_4 = 2.9592$, $d_4 = -0.036$

5th iteration: $a_5 = 0.9845$, $b_5 = 1.9848$, $c_5 = 2.9851$, $d_5 = -0.0158$

6th iteration: $a_6 = 0.9939$, $b_6 = 1.9938$, $c_6 = 2.9938$, $d_6 = -0.006$

7th iteration: $a_7 = 0.99975$, $b_7 = 1.9975$, $c_7 = 2.9976$, $d_7 = -0.0025$

8th iteration: $a_8 = 0.9990$, $b_8 = 1.9990$, $c_8 = 2.9990$, $d_8 = -0.0010$

9th iteration: $a_9 = 0.9996$, $b_9 = 1.9996$, $c_9 = 2.9996$, $d_9 = -0.0004$

10th iteration: $a_{10} = 0.9998$, $b_{10} = 1.9998$, $c_{10} = 2.9998$, $d_{10} = -0.0002$

11th iteration: $a_{11} = 0.9999$, $b_{11} = 1.9999$, $c_{11} = 2.9999$, $d_{11} = -0.0001$

12th iteration: $a_{12} = 1.0000$, $b_{12} = 2.0000$, $c_{12} = 3.0000$, $d_{12} = 0.0000$

From iteration 11th and 12th, it is clear that the difference between these iterations is approximately same, so need to stop the iteration. Hence the solution of system of equations is

$a = 1$, $b = 2$, $c = 3$, $d = 0$

2.12 Gauss-Seidal Iteration Method

It is iterative method in order to find the solution of system of linear equations. This method is also called as '**method of successive displacements**'. Consider the

$$a_{11}x_1 + a_{12}x_2 + a_{13}x_3 + \cdots + a_{1n}x_n = b_1$$
$$a_{11}x_1 + a_{12}x_2 + a_{13}x_3 + \cdots + a_{1n}x_n = b_1$$
$$a_{11}x_1 + a_{12}x_2 + a_{13}x_3 + \cdots + a_{1n}x_n = b_1$$
$$\vdots \quad \vdots \quad \vdots \quad \vdots \quad \vdots \quad \vdots$$
$$a_{n1}x_1 + a_{n2}x_2 + a_{n3}x_3 + \cdots + a_{nn}x_n = b_n$$

Assumption, $a_{ii} \neq 0$ for $i = 1, 2, 3, \ldots.., n$. If this condition does not satisfied, then need to re-arrange the equations to fulfill the criteria.

The above equations can be re-written as

$$x_1 = \frac{b_1}{a_{11}} - \frac{a_{12}}{a_{11}}x_2 - \frac{a_{13}}{a_{11}}x_3 - \cdots - \frac{a_{1n}}{a_{11}}x_n$$
$$x_2 = \frac{b_2}{a_{22}} - \frac{a_{21}}{a_{22}}x_1 - \frac{a_{23}}{a_{22}}x_3 - \cdots - \frac{a_{2n}}{a_{22}}x_n$$
$$x_3 = \frac{b_3}{a_{33}} - \frac{a_{31}}{a_{33}}x_1 - \frac{a_{32}}{a_{33}}x_2 - \cdots - \frac{a_{3n}}{a_{33}}x_n$$
$$\vdots \quad \vdots \quad \vdots \quad \vdots \quad \vdots \quad \vdots$$
$$x_n = \frac{b_n}{a_{nn}} - \frac{a_{n1}}{a_{nn}}x_1 - \frac{a_{n2}}{a_{nn}}x_2 - \cdots - \frac{a_{n,n-1}}{a_{nn}}x_{n-1}$$

Start a solution with an approximation i.e., $x_1^{(0)}, x_2^{(0)}, x_3^{(0)}, \ldots\ldots.. x_n^{(0)}$ are the initial approximate values of unknown variables $x_1, x_2, x_3, \ldots\ldots.. x_n$

wherere $x_1^{(0)}$ presents $x_1 = 0$ and so on.

Note: It is not necessary to take the initial values as zero, it may be reasonable guess.

Substitute the initial values in the above set of system equations and it becomes 1st approximation as

$$x_1^{(1)} = \frac{b_1}{a_{11}} - \frac{a_{12}}{a_{11}}x_2^{(0)} - \frac{a_{13}}{a_{11}}x_3^{(0)} - \cdots - \frac{a_{1n}}{a_{11}}x_n^{(0)}$$

At this stage, the new value of is put inand other variables remain the same, as

$$x_2^{(1)} = \frac{b_2}{a_{22}} - \frac{a_{21}}{a_{22}} x_1^{(1)} - \frac{a_{23}}{a_{22}} x_3^{(0)} - \cdots - \frac{a_{2n}}{a_{22}} x_n^{(0)}$$

Similarly, the new value of andare put inand other variables remain the same, as

$$x_3^{(1)} = \frac{b_3}{a_{33}} - \frac{a_{31}}{a_{33}} x_1^{(1)} - \frac{a_{32}}{a_{33}} x_2^{(1)} - \cdots - \frac{a_{3n}}{a_{33}} x_n^{(0)}$$

and so on $\vdots \quad \vdots \quad \vdots \quad \vdots \quad \vdots$

$$x_n^{(1)} = \frac{b_n}{a_{nn}} - \frac{a_{n1}}{a_{nn}} x_1^{(1)} - \frac{a_{n2}}{a_{nn}} x_2^{(1)} - \cdots - \frac{a_{n,n-1}}{a_{nn}} x_{n-1}^{(1)}$$

Then 2nd iteration / approximation; $x_1^{(1)}, x_2^{(1)}, x_3^{(1)}, \ldots\ldots\ldots x_n^{(1)}$ are the next values of unknown variables $x_1, x_2, x_3, \ldots\ldots\ldots\ldots x_n$

Substitute these 1st iteration values in the above set of system equations and it becomes 2nd approximation as

$$x_1^{(2)} = \frac{b_1}{a_{11}} - \frac{a_{12}}{a_{11}} x_2^{(1)} - \frac{a_{13}}{a_{11}} x_3^{(1)} - \cdots - \frac{a_{1n}}{a_{11}} x_n^{(1)}$$

$$x_2^{(2)} = \frac{b_2}{a_{22}} - \frac{a_{21}}{a_{22}} x_1^{(2)} - \frac{a_{23}}{a_{22}} x_3^{(1)} - \cdots - \frac{a_{2n}}{a_{22}} x_n^{(1)}$$

$$x_3^{(2)} = \frac{b_3}{a_{33}} - \frac{a_{31}}{a_{33}} x_1^{(2)} - \frac{a_{32}}{a_{33}} x_2^{(2)} - \cdots - \frac{a_{3n}}{a_{33}} x_n^{(1)}$$

$$\vdots \quad \vdots \quad \vdots \quad \vdots \quad \vdots \quad \vdots$$

$$x_1^{(2)} = \frac{b_n}{a_{nn}} - \frac{a_{n1}}{a_{nn}} x_1^{(2)} - \frac{a_{n2}}{a_{nn}} x_2^{(2)} - \cdots - \frac{a_{n,n-1}}{a_{nn}} x_{n-1}^{(2)}$$

Then proceed to 3rd & 4th approximation and so on in order to get the exact solution. It must ensure that the difference between the two successive values of unknown variables should be approximately same.

Example 2.15: Solve the system of equations 10a – 2b – c – d = 3; – 2a + 10b – c – d = 15; – a – b + 10c – 2d = 27; – a – b – 2c + 10d = – 9, using Gauss-Seidel iteration method.

Solution: The given system of equations is

$10a - 2b - c - d = 3$ (1)

$-2a + 10b - c - d = 15$ (2)

$-a - b + 10c - 2d = 27$ (3)

$-a - b - 2c + 10d = -9$ (4)

Re-write the equations 1, 2, 3 & 4 as

$$a = \frac{1}{10}(3 + 2b + c + d)$$

$$b = \frac{1}{10}(15 + 2a + c + d)$$

$$c = \frac{1}{10}(27 + a + b + 2d)$$

$$d = \frac{1}{10}(-9 + a + b + 2c)$$

Start a solution with an approximation i.e., $a_0 = 0$ (or $a^{(0)} = 0$), $b_0 = 0$, $c_0 = 0$, $d_0 = 0$, are the initial approximate values of unknown variables a, b, c and d.

1st iteration:

$$a_1 = \frac{1}{10}(3 + 2b_0 + c_0 + d_0) = \frac{1}{10}(3 + 2(0) + 0 + 0) = 0.3$$

$$b_1 = \frac{1}{10}(15 + 2a_1 + c_0 + d_0) = \frac{1}{10}(15 + 2(0.3) + 0 + 0) = 1.56$$

$$c_1 = \frac{1}{10}(27 + a_1 + b_1 + 2d_0) = \frac{1}{10}(27 + 0.3 + 1.56 + 2(0)) = 2.886$$

$$d_1 = \frac{1}{10}(-9 + a_1 + b_1 + 2c_1) = \frac{1}{10}(-9 + 0.3 + 1.56 + 2(2.886)) = -0.1368$$

2nd iteration:

$$a_2 = \frac{1}{10}(3 + 2b_1 + c_1 + d_1) = \frac{1}{10}(3 + 2(1.56) + 2.886 + (-0.1368)) = 0.8869$$

$$b_2 = \frac{1}{10}(15 + 2a_2 + c_1 + d_1) = \frac{1}{10}(15 + 2(0.8869) + 2.886 + (-0.1368)) = 1.9523$$

$$c_2 = \frac{1}{10}(27 + a_2 + b_2 + 2d_1) = \frac{1}{10}(27 + 0.8869 + 1.9523 + 2(-0.1368)) = 2.9566$$

$$d_2 = \frac{1}{10}(-9 + a_2 + b_2 + 2c_2) = \frac{1}{10}(-9 + 0.8869 + 1.9523 + 2\,(2.9566)) = -0.0248$$

3rd iteration:

$$a_3 = \frac{1}{10}(3 + 2b_2 + c_2 + d_2) = \frac{1}{10}(3 + 2\,(1.9523) + 2.9566 + (-0.0248)) = 0.9836$$

$$b_3 = \frac{1}{10}(15 + 2a_3 + c_2 + d_2) = \frac{1}{10}(15 + 2\,(0.9836) + 2.9566 + (-0.0248)) = 1.9899$$

$$c_3 = \frac{1}{10}(27 + a_3 + b_3 + 2d_2) = \frac{1}{10}(27 + 0.9836 + 1.9899 + 2\,(-0.0248)) = 2.9924$$

$$d_3 = \frac{1}{10}(-9 + a_3 + b_3 + 2c_3) = \frac{1}{10}(-9 + 0.9836 + 1.9899 + 2\,(2.9924)) = -0.0042$$

4th iteration:

$$a_4 = \frac{1}{10}(3 + 2b_3 + c_3 + d_3) = \frac{1}{10}(3 + 2\,(1.9899) + 2.9924 + (-0.0042)) = 0.9968$$

$$b_4 = \frac{1}{10}(15 + 2a_4 + c_3 + d_3) = \frac{1}{10}(15 + 2\,(0.9968) + 2.9924 + (-0.0042)) = 1.9982$$

$$c_4 = \frac{1}{10}(27 + a_4 + b_4 + 2d_3) = \frac{1}{10}(27 + 0.9968 + 1.9982 + 2\,(-0.0042)) = 2.9987$$

$$d_4 = \frac{1}{10}(-9 + a_4 + b_4 + 2c_4) = \frac{1}{10}(-9 + 0.9968 + 1.9982 + 2\,(2.9987)) = -0.0008$$

5th iteration:

$$a_5 = \frac{1}{10}(3 + 2b_4 + c_4 + d_4) = \frac{1}{10}(3 + 2\,(1.9982) + 2.9987 + (-0.0008)) = 0.9994$$

$$b_5 = \frac{1}{10}(15 + 2a_5 + c_4 + d_4) = \frac{1}{10}(15 + 2\,(0.9994) + 2.9987 + (-0.0008)) = 1.9997$$

$$c_5 = \frac{1}{10}(27 + a_5 + b_5 + 2d_4) = \frac{1}{10}(27 + 0.9994 + 1.9997 + 2\,(-0.0008)) = 2.9998$$

$$d_5 = \frac{1}{10}(-9 + a_3 + b_3 + 2c_3) = \frac{1}{10}(-9 + 0.9994 + 1.9997 + 2\,(2.9998)) = -0.0001$$

6th iteration:

$$a_6 = \frac{1}{10}(3 + 2b_5 + c_5 + d_5) = \frac{1}{10}(3 + 2\,(1.9997) + 2.9998 + (-0.0001)) = 0.9999$$

$$b_6 = \frac{1}{10}(15 + 2a_6 + c_5 + d_5) = \frac{1}{10}(15 + 2\,(0.9999) + 2.9998 + (-0.0001)) = 1.9999$$

$$c_6 = \frac{1}{10}(27 + a_6 + b_6 + 2d_5) = \frac{1}{10}(27 + 0.9999 + 1.9999 + 2\,(-0.0001)) = 3.0000$$

$$d_6 = \frac{1}{10}(-9 + a_6 + b_6 + 2c_6) = \frac{1}{10}\,(-9 + 0.9999 + 1.9999 + 2\,(3.0000)) = 0.0000$$

7th iteration:

$$a_7 = \frac{1}{10}(3 + 2b_6 + c_6 + d_6) = \frac{1}{10}(3 + 2\,(1.9999) + 3.0000 + (0.0000)) = 1.0000$$

$$b_7 = \frac{1}{10}(15 + 2a_7 + c_6 + d_6) = \frac{1}{10}(15 + 2\,(1.0000) + 3.0000 + (0.0000)) = 2.0000$$

$$c_7 = \frac{1}{10}(27 + a_7 + b_7 + 2d_6) = \frac{1}{10}(27 + 1.0000 + 2.0000 + 2\,(0.0000)) = 3.0000$$

$$d_7 = \frac{1}{10}(-9 + a_7 + b_7 + 2c_7) = \frac{1}{10}(-9 + 1.0000 + 2.0000 + 2\,(3.0000)) = 0.0000$$

From iteration 6th and 7th, it is clear that the difference between these iterations is approximately same, so need to stop the iteration. Hence the solution of system of equations is

$a = 1, b = 2, c = 3, d = 0$

2.13 Iterative Method for Eigen Values and Eigen Vectors

Consider square matrix $A = [a_{ij}]$ of nth order. Solution yields a constant λ and a column matrix X, i.e.,

$$A\,X = \lambda\,X \qquad (1)$$

$$\Rightarrow (A - \lambda\,I)\,X = 0 \qquad (2)$$

$$\text{where } X = \begin{bmatrix} x_1 \\ x_2 \\ \vdots \\ x_n \end{bmatrix}$$

The above form of matrix equation represents n homogeneous linear equations

$$(a_{11} - \lambda)\,x_1 + a_{12}\,x_2 + \dots\dots\dots + a_{1n}\,x_n = 0$$

$$a_{21}\,x_1 + (a_{22} - \lambda)\,x_2 + \dots\dots\dots + a_{2n}\,x_n = 0$$

....

$a_{n1}x_1 + a_{n2}x_2 + \ldots\ldots\ldots + (a_{nn} - \lambda)x_n = 0$

This will have a non-trivial solution only if the coefficients determinant vanishes i.e.,

$$\begin{vmatrix} a_{11}-\lambda & a_{12} & \cdots & a_{1n} \\ a_{21} & a_{21}-\lambda & \cdots & a_{2n} \\ \vdots & \vdots & \vdots & \vdots \\ a_{n1} & a_{n2} & \cdots & a_{nn}-\lambda \end{vmatrix} = 0 \qquad (3)$$

The equation (3) gives an nth degree equation in λ, called the characteristic equation of the matrix A. Its roots λ_i (i = 1, 2, 3, ... n) are called eigen value (or characteristic root / latent root) and corresponding to each eigen value, the equation (2) have a non-zero solution, so this column matrix X is called corresponding eigen vector (or characteristic vector / latent vector / pole / invariant vector). Thus $|A - \lambda I| = 0$ is called characteristic equation of square matrix A of degree n in λ and determinant $|A - \lambda I|$ is called characteristic polynomial of square matrix A. The above such equation can be solved easily, but for longer system of equations, it is better to use other better methods.

So in order to find the eigen values and its corresponding eigen vectors, two methods i.e., Jacobi's method and Reyleigh's Power method are discussed.

2.13.1 Jacobi's Method

This method can be applied for real symmetric matrix, A, and its eigen values are real and hence exist a real orthodiagonal matrix T such that $T^{-1}AT$, which is a diagonal matrix D. So in this method, diagonalising A through the application of series of orthodiagonal transformations T_1, T_2, ..., such that their product satisfies the equation $T^{-1}AT=D$.

Let's select the numerically largest non-diagonal element a_{ij} and form a 2 X 2 sub matrix A_1 as, $\begin{bmatrix} a_{ii} & a_{ij} \\ a_{ji} & a_{jj} \end{bmatrix}$ where $a_{ij} = a_{ji}$ which can be easily diagonalised.

Suppose T_1 is an orthodiagonal matrix $= \begin{bmatrix} \cos\theta & -\sin\theta \\ \sin\theta & \cos\theta \end{bmatrix}$, such that $T_1^{-1} = T_1'$

Then $T_1^{-1} A_1 T_1 = \begin{bmatrix} \cos\theta & \sin\theta \\ -\sin\theta & \cos\theta \end{bmatrix} \begin{bmatrix} a_{ii} & a_{ij} \\ a_{ji} & a_{jj} \end{bmatrix} \begin{bmatrix} \cos\theta & -\sin\theta \\ \sin\theta & \cos\theta \end{bmatrix}$

$$= \begin{bmatrix} a_{ii}\cos^2\theta + a_{jj}\sin^2\theta + a_{ij}\sin 2\theta & a_{ij}\cos 2\theta + \frac{1}{2}(a_{jj} - a_{ii})\sin 2\theta \\ a_{ij}\cos 2\theta + \frac{1}{2}(a_{jj} - a_{ii})\sin 2\theta & a_{ii}\sin^2\theta + a_{jj}\cos^2\theta - a_{ij}\sin 2\theta \end{bmatrix}$$

This above can be reduced to diagonal matrix form if the element, $a_{ij}\cos 2\theta + \frac{1}{2}(a_{jj} - a_{ii})\sin 2\theta$, reduces to zero, if $\tan 2\theta = \frac{2a_{ij}}{a_{ii} - a_{jj}}$, through this 4 values are yielded and for this selection is in between $-\frac{\pi}{4} \le \theta \le \frac{\pi}{4}$. This results in reduction to a diagonal matrix.

Next step is to find the largest non-diagonal element in the new rotated matrix and then repeat the above procedure with the help of orthodiagonal matrix T_2. In this way, a new series of such transformations are performed so as to annihilate the non-diagonal elements. After making n transformations, result is $T_n^{-1} T_{n-1}^{-1} \ldots T_1^{-1} AT_1 \ldots T_2^{-1} A_2 T_2 \ldots T_{n-1}^{-1} T_n = T^{-1} A T$, as $n \to \infty$,

$T^{-1} A T$ approaches a diagonal matrix whose diagonal elements are the eigen values of A.

A drawback of this method is that the elements annihilated by a transformation, may not remain zero during the subsequent transformations.

Example 2.16: Using Jacobi's method, find all the eigen values and the eigen vectors of the matrix.

$$A = \begin{bmatrix} 2 & \sqrt{2} & 0 \\ \sqrt{2} & 3 & \sqrt{2} \\ 2 & \sqrt{2} & 1 \end{bmatrix}$$

Solution: The largest non-diagonal (or off diagonal) element is $a_{13} = a_{31} = 2$, as $a_{11} = a_{33} = 1$.

$$\tan 2\theta = \frac{2a_{13}}{a_{11} - a_{33}} = \frac{2 X 2}{1-1} = \frac{4}{0} = \infty$$

$$\Rightarrow 2\theta = \frac{\pi}{2} \quad \Rightarrow \quad \theta = \frac{\pi}{4}$$

The transformation matrix, $T_1 = \begin{bmatrix} \cos\theta & 0 & -\sin\theta \\ 0 & 1 & 0 \\ \sin\theta & 0 & \cos\theta \end{bmatrix} = \begin{bmatrix} \frac{1}{\sqrt{2}} & 0 & -\frac{1}{\sqrt{2}} \\ 0 & 1 & 0 \\ \frac{1}{\sqrt{2}} & 0 & \frac{1}{\sqrt{2}} \end{bmatrix}$

Inverse of the transformation matrix, T_1^{-1} (= T_1'; because T is orthogonal matrix)

$$= \begin{bmatrix} \frac{1}{\sqrt{2}} & 0 & \frac{1}{\sqrt{2}} \\ 0 & 1 & 0 \\ -\frac{1}{\sqrt{2}} & 0 & \frac{1}{\sqrt{2}} \end{bmatrix}$$

The 1st transformation gives, $A_1 = T_1^{-1} A\, T_1$

$$A_1 = \begin{bmatrix} 2 & \sqrt{2} & 0 \\ \sqrt{2} & 3 & \sqrt{2} \\ 2 & \sqrt{2} & 1 \end{bmatrix} \begin{bmatrix} \frac{1}{\sqrt{2}} & 0 & \frac{1}{\sqrt{2}} \\ 0 & 1 & 0 \\ -\frac{1}{\sqrt{2}} & 0 & \frac{1}{\sqrt{2}} \end{bmatrix} \begin{bmatrix} \frac{1}{\sqrt{2}} & 0 & -\frac{1}{\sqrt{2}} \\ 0 & 1 & 0 \\ \frac{1}{\sqrt{2}} & 0 & \frac{1}{\sqrt{2}} \end{bmatrix}$$

$$A_1 = \begin{bmatrix} 3 & 2 & 0 \\ 2 & 3 & 0 \\ 0 & 2 & -1 \end{bmatrix}$$

So in A_1, the largest non-diagonal (or off diagonal) element is $a_{12} = a_{21} = 2$, and $a_{11} = a_{22} = 3$.

$$\tan 2\theta = \frac{2a_{12}}{a_{11} - a_{22}} = \frac{2 X 2}{1-1} = \frac{4}{0} = \infty$$

$$\Rightarrow 2\theta = \frac{\pi}{2} \quad \Rightarrow \quad \theta = \frac{\pi}{4}$$

The second transformation matrix, $T_2 = \begin{bmatrix} \cos\theta & -\sin\theta & 0 \\ \sin\theta & \cos\theta & 0 \\ 0 & 0 & 1 \end{bmatrix} = \begin{bmatrix} \frac{1}{\sqrt{2}} & -\frac{1}{\sqrt{2}} & 0 \\ \frac{1}{\sqrt{2}} & \frac{1}{\sqrt{2}} & 0 \\ 0 & 0 & 1 \end{bmatrix}$

Inverse of the transformation matrix T_2 is $T_2^{-1} = \begin{bmatrix} \frac{1}{\sqrt{2}} & \frac{1}{\sqrt{2}} & 0 \\ -\frac{1}{\sqrt{2}} & \frac{1}{\sqrt{2}} & 0 \\ 0 & 0 & 1 \end{bmatrix}$

The 2nd transformation gives, $A_1 = T_2^{-1} A_1 T_2$

$$A_2 = \begin{bmatrix} \frac{1}{\sqrt{2}} & \frac{1}{\sqrt{2}} & 0 \\ -\frac{1}{\sqrt{2}} & \frac{1}{\sqrt{2}} & 0 \\ 0 & 0 & 1 \end{bmatrix} \begin{bmatrix} 3 & 2 & 0 \\ 2 & 3 & 0 \\ 0 & 2 & -1 \end{bmatrix} \begin{bmatrix} \frac{1}{\sqrt{2}} & -\frac{1}{\sqrt{2}} & 0 \\ \frac{1}{\sqrt{2}} & \frac{1}{\sqrt{2}} & 0 \\ 0 & 0 & 1 \end{bmatrix}$$

$$A_2 = \begin{bmatrix} 5 & 0 & 0 \\ 0 & 1 & 0 \\ 0 & 0 & -1 \end{bmatrix}$$

Hence the eigen values of the given matrix are 5, 1, –1 and the corresponding eigen vectors are the columns of

$$T = T_1 T_2 = \begin{bmatrix} \frac{1}{\sqrt{2}} & 0 & -\frac{1}{\sqrt{2}} \\ 0 & 1 & 0 \\ \frac{1}{\sqrt{2}} & 0 & \frac{1}{\sqrt{2}} \end{bmatrix} \begin{bmatrix} \frac{1}{\sqrt{2}} & -\frac{1}{\sqrt{2}} & 0 \\ \frac{1}{\sqrt{2}} & \frac{1}{\sqrt{2}} & 0 \\ 0 & 0 & 1 \end{bmatrix} = \begin{bmatrix} \frac{1}{2} & -\frac{1}{\sqrt{2}} & -\frac{1}{\sqrt{2}} \\ \frac{1}{\sqrt{2}} & \frac{1}{\sqrt{2}} & 0 \\ \frac{1}{2} & -\frac{1}{\sqrt{2}} & \frac{1}{\sqrt{2}} \end{bmatrix}$$

2.13.2 Reyleigh's Power Method

This method is suitable for machine computations to determine numerically the largest eigen value and the corresponding eigen vector.

If $X_1, X_2, X_3 \ldots\ldots X_n$ are the eigen vectors corresponding to the eigen values $\lambda_1, \lambda_2, \lambda_3 \ldots\ldots \lambda_n$ then the column vector can be written as

$$X = k_1 X_1 + k_2 X_2 + k_3 X_3 + \ldots\ldots. k_n X_n$$

Then

$$A X = k_1 A X_1 + k_2 A X_2 + k_3 A X_3 + \ldots\ldots. k_n A X_n$$

$$A X = k_1 \lambda_1 X_1 + k_2 \lambda_2 X_2 + k_3 \lambda_3 X_3 + \ldots\ldots. k_n \lambda_n X_n$$

Then

$$A^2 X = k_1 \lambda_1^2 X_1 + k_2 \lambda_2^2 X_2 + k_3 \lambda_3^2 X_3 + \ldots\ldots. k_n \lambda_n^2 X_n$$

So on

$$A^r X = k_1 \lambda_1^r X_1 + k_2 \lambda_2^r X_2 + k_3 \lambda_3^r X_3 + \ldots\ldots. k_n \lambda_n^r X_n$$

If $|\lambda_1| > |\lambda_2| > |\lambda_3| \ldots\ldots. |\lambda_n|$, then λ_1 is the largest root and the efforts of $k_1 \lambda_1^r X_1$ to the sum on the right increases with r, so on multiplying a column vector by A results in value nearer to the eigen vector X_1.Then make the largest component of the resulting vector unity in order to avoid k_1.

Assume any guessed value of eigen vector initially, otherwise take $\begin{bmatrix} 1 \\ 0 \\ 0 \end{bmatrix}$ as $X^{(0)}$.

Multiply the given matrix with this eigen vector. The maximum value in the outcome matrix is taken out (called λ) and the resulting eigen vector is the 1st approximation as $X^{(1)}$. Then for 2nd approximation, the given matrix is multiplied with $X^{(1)}$ to get $X^{(2)}$. This process is repeated until $[X^r - X^{r-1}]$ becomes very small, then λ^r is the largest eigen value and X^r is the corresponding eigen vector.

Example 2.17: Find the largest eigen value and the corresponding eigen vector of the matrix.

$$A = \begin{bmatrix} 2 & -1 & 0 \\ -1 & 2 & -1 \\ 0 & -1 & 2 \end{bmatrix}$$

Solution: Let the initial eigen vector is $X^{(0)} = \begin{bmatrix} 1 \\ 0 \\ 0 \end{bmatrix}$

Then $A\,X^{(0)} = \begin{bmatrix} 2 & -1 & 0 \\ -1 & 2 & -1 \\ 0 & -1 & 2 \end{bmatrix}\begin{bmatrix} 1 \\ 0 \\ 0 \end{bmatrix} = \begin{bmatrix} 2 \\ -1 \\ 0 \end{bmatrix} = 2\begin{bmatrix} 1 \\ -0.5 \\ 0 \end{bmatrix}$

$$A\,X^{(0)} = 2\begin{bmatrix} 1 \\ -0.5 \\ 0 \end{bmatrix} = \lambda_1 X^{(1)}$$

$$A\,X^{(1)} = \begin{bmatrix} 2 & -1 & 0 \\ -1 & 2 & -1 \\ 0 & -1 & 2 \end{bmatrix}\begin{bmatrix} 1 \\ -0.5 \\ 0 \end{bmatrix} = \begin{bmatrix} 2.5 \\ -2 \\ 1.2 \end{bmatrix} = 2.5\begin{bmatrix} 1 \\ -0.8 \\ 0.2 \end{bmatrix}$$

$$A\,X^{(1)} = 2.5\begin{bmatrix} 1 \\ -0.8 \\ 0.2 \end{bmatrix} = \lambda_2 X^{(2)}$$

$$A\,X^{(2)} = \begin{bmatrix} 2 & -1 & 0 \\ -1 & 2 & -1 \\ 0 & -1 & 2 \end{bmatrix}\begin{bmatrix} 1 \\ -0.8 \\ 0.2 \end{bmatrix} = \begin{bmatrix} 2.8 \\ -2.8 \\ 1.2 \end{bmatrix} = 2.8\begin{bmatrix} 1 \\ -1 \\ 0.42857 \end{bmatrix}$$

$$A\,X^{(2)} = 2.8\begin{bmatrix} 1 \\ -1 \\ 0.42857 \end{bmatrix} = \lambda_3 X^{(3)}$$

$$A\,X^{(3)} = \begin{bmatrix} 2 & -1 & 0 \\ -1 & 2 & -1 \\ 0 & -1 & 2 \end{bmatrix}\begin{bmatrix} 1 \\ -1 \\ 0.42857 \end{bmatrix} = \begin{bmatrix} 3 \\ -3.4286 \\ 1.8571 \end{bmatrix} = 3.4286\begin{bmatrix} 0.87499 \\ -1 \\ 0.54165 \end{bmatrix}$$

$$A\,X^{(3)} = 3.4286\begin{bmatrix} 0.87499 \\ -1 \\ 0.54165 \end{bmatrix} = \lambda_4 X^{(4)}$$

$$A\,X^{(4)} = \begin{bmatrix} 2 & -1 & 0 \\ -1 & 2 & -1 \\ 0 & -1 & 2 \end{bmatrix} \begin{bmatrix} 0.87499 \\ -1 \\ 0.54165 \end{bmatrix} = \begin{bmatrix} 2.75 \\ -3.4166 \\ 2.0833 \end{bmatrix} = 3.4166 \begin{bmatrix} 0.80489 \\ -1 \\ 0.60976 \end{bmatrix}$$

$$A\,X^{(4)} = 3.4166 \begin{bmatrix} 0.80489 \\ -1 \\ 0.60976 \end{bmatrix} = \lambda_5\, X^{(5)}$$

$$A\,X^{(5)} = \begin{bmatrix} 2 & -1 & 0 \\ -1 & 2 & -1 \\ 0 & -1 & 2 \end{bmatrix} \begin{bmatrix} 0.80489 \\ -1 \\ 0.60976 \end{bmatrix} = \begin{bmatrix} 2.6098 \\ -3.4147 \\ 2.2195 \end{bmatrix} = 3.4147 \begin{bmatrix} 0.76402 \\ -1 \\ 0.64998 \end{bmatrix}$$

$$A\,X^{(5)} = 3.4147 \begin{bmatrix} 0.76402 \\ -1 \\ 0.64998 \end{bmatrix} = \lambda_6\, X^{(6)}$$

$$A\,X^{(6)} = \begin{bmatrix} 2 & -1 & 0 \\ -1 & 2 & -1 \\ 0 & -1 & 2 \end{bmatrix} \begin{bmatrix} 0.76402 \\ -1 \\ 0.64998 \end{bmatrix} = \begin{bmatrix} 2.5280 \\ -3.4140 \\ 2.3000 \end{bmatrix} = 3.4140 \begin{bmatrix} 0.74048 \\ -1 \\ 0.67370 \end{bmatrix}$$

$$A\,X^{(6)} = 3.4140 \begin{bmatrix} 0.74048 \\ -1 \\ 0.67370 \end{bmatrix} = \lambda_7\, X^{(7)}$$

Hence the largest eigen value is 3.4140 and its corresponding eigen vector is

$$\begin{bmatrix} 0.74048 \\ -1 \\ 0.67370 \end{bmatrix}$$

Exercise

2.1 What is meant by simultaneous equations?

2.2 Write a note on the matrices, its fields and operations related to matrices.

2.3 How linear equations can be solved?

2.4 Determine the inverse of, $\begin{bmatrix} -1 & 1 & 5 \\ 3 & -1 & 4 \\ 2 & 3 & 6 \end{bmatrix}, \begin{bmatrix} 10 & 3 & 10 \\ 8 & -2 & 9 \\ 8 & 1 & -10 \end{bmatrix}$ and $\begin{bmatrix} 5 & 3 \\ 3 & 2 \end{bmatrix}$

2.5 Solve the linear equations through Cramer's rule

(a) $2x + 3y - z = -10$

$-x + 4y + 2z = -4$

$2x - 2y + 5z = 35$

(b) $x_1 + 2x_2 + 3x_3 + 4x_4 = 8$

$2x_1 - 2x_2 - x_3 - x_4 = -3$

$x_1 - 3x_2 + 4x_3 - 4x_4 = 8$

$2x_1 + 2x_2 - 3x_3 + 4x_4 = -2$

2.6 Solve the following set of simultaneous linear equations through matrix inversion method

(a) $2x + 3y - z = -10$

$-x + 4y + 2z = -4$

$2x - 2y + 5z = 35$

(b) $3x + 2y = 10$

$-5x + 10y = 20$

(c) $x_1 + 2x_2 + 3x_3 + 4x_4 = 8$

$2x_1 - 2x_2 - x_3 - x_4 = -3$

$x_1 - 3x_2 + 4x_3 - 4x_4 = 8$

$2x_1 + 2x_2 - 3x_3 + 4x_4 = -2$

2.7 Solve the following problems using Gauss Jordan and Gauss elimination method

(a) $x_1 + 2x_2 + 3x_3 + 4x_4 = 8$

$2x_1 - 2x_2 - x_3 - x_4 = -3$

$x_1 - 3x_2 + 4x_3 - 4x_4 = 8$

$2x_1 + 2x_2 - 3x_3 + 4x_4 = -2$

(b) $6x + 6y + 3z = 17$

$x + 2y + 2z = 11$

(c) $2a + b + c = 4$

$3b - 3c = 0$

$-b + 2c = 1$

(d) $x - y = 2$

$-2x + 2y - z = -1$

$y - 2z = 6$

(e) $4a - 2b - 3c + 6d = 12$

$-5a + 7b + 6.5c - 6d = -6.5$

$a + 7.5b + 6.25c + 5.5d = 16$

$-12a + 22b + 15c - d = 17$

2.8 Solve the simultaneous linear equation through Crout's method.

(a) $x + 0.5y = 1$

$0.5x + y + 0.5z = 1$

$0.5y + z = 3$

(b) $20x + y - 2z = 17$

$3x + 20y - z = -18$

$2x - 3y + 20z = 25$

2.9 Reduce the matrix $\begin{bmatrix} 1 & 1 & 0.5 \\ 1 & 1 & 0.25 \\ 0.5 & 0.25 & 2 \end{bmatrix}$ into the tri-diagonal form through Choleski's method.

2.10 Solve the simultaneous linear equation through Thomas Algorithm method

(a) $a + 2b + 3c + 5d = 10$

$2a - 3b - 2c - d = -9$

$a - b + c - 4d = 12$

$2a + 4b - 6c + 4d = -3$

(b) $2x_1 + x_2 = 3$

$x_1 + 3x_2 + x_3 = 3$

$x_2 + x_3 + 2x_4 = 3$

$2x_3 + 3x_4 = 4$

2.11 Solve the system of equations $2x + 4y + 10z = 30$; $x - 4y - z = 18$; $-x + 4y + 7z = 35$ through Jacobi's iteration method.

2.12 Solve the following system of equations using the Gauss-Seidal method.

(a) $10x_1 - 2x_2 - x_3 - x_4 = 3$

$-2x_1 + 10x_2 - x_3 - x_4 = 15$

$-x_1 - x_2 + 10x_3 - 2x_4 = 27$

$-x_1 - x_2 - 2x_3 + 10x_4 = -9$

(b) $2x - y + 5z = 15$

$2x + y + z = 7$

$2x + 3y + z = 10$

2.13 Using Jacobi's method, find all the eigen values and the eigen vectors of the matrices

$$\text{(a)} \begin{bmatrix} 5 & 0 & 1 \\ 0 & -2 & 0 \\ 1 & 0 & 5 \end{bmatrix} \text{(b)} \begin{bmatrix} 2 & 3 & 1 \\ 3 & 2 & 2 \\ 1 & 2 & 1 \end{bmatrix}$$

2.14 Through Reyleigh's power method, find the largest eigen value and the corresponding eigen vector of the matrix

$$\text{(a)} \begin{bmatrix} -2 & 1 & 0 \\ -5 & 2 & 7 \\ 10 & -1 & 2 \end{bmatrix} \text{(b)} \begin{bmatrix} 3 & 1 & 1 \\ 9 & -5 & 15 \\ 8 & -1 & 20 \end{bmatrix}$$

3

Solution of Algebric and Transcendental Equations

3.1 Introduction

An expression of the form $[f_n(x) = a_0x^n + a_1x^{n-1} + a_2x^{n-2} + ... + a_{n-1}x + a_n]$, where a's are constants and n is a positive integer, is said to be a polynomial in x of degree n, provided $a_0 \neq 0$. The values of x for which the equation $f_n(x)$ becomes zero are called zeros or the roots of the polynomial and every polynomial of nth degree that has n zeros.

The equations of the form $f_n(x) = 0$ are called Algebric and Transcendental equations according as $f_n(x)$ is purely a polynomial in x or contains some other functions such as logarithmic, exponential and trigonometric functions etc. The equation, $x^6 - x^4 - x^3 - 1 = 0$, is called an **algebraic equation.** But, if $f_n(x)$ involves trignometrical, arithmetical or exponential terms in it, then it is called **transcendental equation**, for example; $xe^x - 2 = 0$ and $x\log_{10}x - 1.2 = 0$.

Basic Properties of an Algebraic Equation and its Roots:

- If f(x) is exactly divisible by $(x-\alpha)$, then α is a root of f (x).
- Every algebraic equation of nth degree has n and only n real or imaginary roots.

  Conversely, if $\alpha_1, \alpha_2, ... , \alpha_n$; be the n roots of the nth degree equation $f(x) = 0$, then $f(x) = A(x-\alpha_1)(x-\alpha_2).........(x - \alpha_n)$.
- If f(x) is continuous in the interval [a, b] and f(a), f(b) have different signs, then the equation have at least one root between x=a and x=b (known as **Intemediate Value Theorem**)
- In an equation with real coefficients, imaginary roots occur in conjugate pairs, i.e. if $\alpha+i\beta$ is root of f(x)=0, then $\alpha-i\beta$ is also its root. Similarly $\alpha+\beta$ is an irrational root of f(x)=0, then $\alpha-\beta$ is also its root.

3.2 Solution of Algebraic and Transcendental Equations

Let the algebraic equation of nth degree be form $[a_0x^n + a_1x^{n-1} + a_2x^{n-2} + ... + a_{n-1}x + a_n]$, $a_0 \neq 0$. The equation with nth degree has n roots, some of which may be real or complex, simple or multiple. A transcendental equation may have one root or no root or infinite number of roots depending on the form of $f_n(x)$. The methods of finding the roots of $f_n(x) = 0$ are classified as,

1. Direct Methods
2. Numerical Methods.

Direct methods give the exact values of all the roots in a finite number of steps. The real roots of any equation we can obtain by the method of synthetic division method. Some theorems are given below concerned with the real roots of the algebraic equation.

Theorem 3.2.1. Descarte's Rule of Signs

If any algebraic equation $f_n(x) = 0$ then

(i) The number of positive roots cannot exceed the number of change of signs (or variations) from positive to negative and from negative to positive in $f_n(x)$.

(ii) The number of negative roots cannot exceed the number of variation in $f_n(-x)$.

Theorem 3.2.2. If two real quantities a and b be substituted for x in any polynomial $f_n(x)$ and if $f_n(a)$ and $f_n(b)$ are of opposite signs, then at least one or an odd number of real roots of the equation $f_n(x) = 0$ lie between a and b. And if $f_n(a)$ and $f_n(b)$ are of the same sign, then either no real root or an even number of roots of $f_n(x) = 0$, lie in between a and b.

Theorem 3.2.3. Every equation of an odd degree has at least one real root of opposite to that of (a_n/a_0), whereas every equations of an even degree whose last term is negative has at least two real roots, one positive and the other negative.

Theorem 3.2.4. The largest root of the equation $[a_0x^n + a_1x^{n-1} + a_2x^{n-2} + ... + a_{n-1}x + a_n]$, $a_0 \neq 0$ may be obtained approximately by the root of the equation $a_0x+a_1=0$, or by the root of the quadratic equation $a_0x^2 + a_1x + a_2 = 0$ which is larger in absolute value, similarly the smallest root can be obtained approximately by the root of the equation $a_{n-1}\, x^n + a_n = 0$ or by the root of the equation $a_{n-2}\, x^2+a_{n-1}\, x+a_n = 0$ which is smaller is absolute value.

Numerical methods are based on the idea of successive approximations. In these methods, we start with one or two initial approximations to the root and obtain a sequence of approximations x_0, x_1, … x_k which in the limit as $k \to \infty$ converge to the exact root x=a.

There are no direct methods for solving higher degree algebraic equations or transcendental equations. Such equations can be solved by Numerical methods. In these methods, we first find an interval in which the root lies. If a and b are two numbers such that f (a) and f (b) have opposite signs, then a root of f (x) = 0 lies in between a and b. We take a or b or any value in between a or b as first approximation x_1. This is further improved by numerical methods. Here we discuss few important Numerical methods to find a root of $f_n(x) = 0$.

3.3 Bisection (or Bolzano) Method

This method of solving transcendental equations, consists in locating the roots of the equation f(x) = 0 between two numbers say a and b such that f(x) is continuous for $a \leq x \leq b$ and f(a) and f(b) are of opposite signs so that the product f(a) f(b) < 0 i.e., the curve cuts the x-axis between a and b. Then the desired root is approximately, $x_1 = \frac{a+b}{2}$. If $f(x_1) = 0$, then x_1 is a root of f(x) = 0. Otherwise the root lies between either 'a and x_1'or 'x_1 and b', according as $f(x_1)$ is positive or negative. Then we bisect the interval as before and continue the process to improve the result to greater accuracy. From the figure 3.1, $f(x_1)$ is positive, so that the root lies between a and x_1 i.e., $x_2 = \frac{a+x_1}{2}$. If $f(x_2)$ is negative, the root lies between x_1 and x_2. Then the next approximation will be $x_3 = \frac{x_1+x_2}{2}$ and so on.

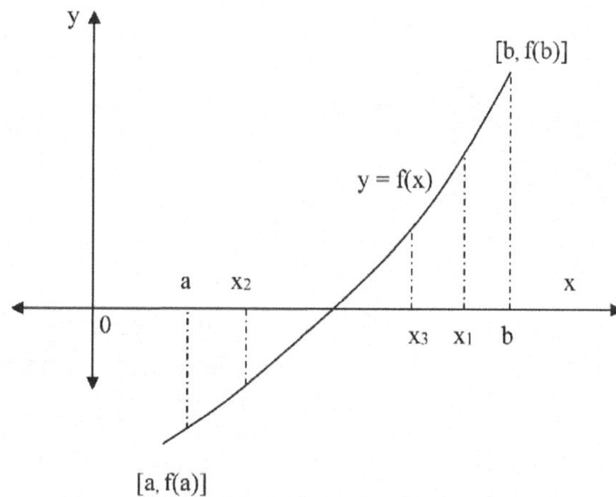

Fig. 3.1: Bisection method

The Bisection Method is also called **'Binary Chopping' or 'Incremental Search Method' or 'Interval Halving method'.**

Working Rule:

Step 1: Choose a lower end as x_0 (x = a) and upper end as x_n (x = b) to guess the roots lies in between them and for that root the sign of the function will change over the interval. This can be checked by ensuring that $f(x_0)\, f(x_n) < 0$.

Step 2: Make the following root x_r is determined by $x_r = \dfrac{x_0 + x_n}{2}$

Step 3: Make the following evaluations to determine in which subinterval the root lies.

(a) If $f(x_0)\, f(x_n) < 0$ root lies in the lower subinterval and repeat the process.

(b) If $f(x_0)\, f(x_n) > 0$ the root lies in upper subinterval and repeat the process.

(c) If $f(x_0)\, f(x_n) = 0$, the root equals x; terminate the computation.

Example 3.1: Solve $x^3 - 9x + 1 = 0$ for the root lies between x = 2 and x = 4 by using Bisection method.

Solution: Here $f(x) = x^3 - 9x + 1$ so that f (x) is continuous in the interval $2 \leq x \leq 4$.

Further, $\begin{cases} f(2) = -9 \\ f(4) = 29 \end{cases}$

and f(2) * f(4) < 0 so that a root lies between x = 2 and x = 4.

Now since $x_r = \dfrac{x_0 + x_n}{2}$, therefore $x_1 = \dfrac{x_0 + x_n}{2} = \dfrac{2+4}{2} = 3$, f(3) =1 and f(2).f(3)<0.

Hence, $x_2 = \dfrac{x_0 + x_1}{2} = \dfrac{2+3}{2} = 2.5$, f(2.5)= –5.87 and since f(3).f(2.5)<0 therefore the root lies between 2.5 and 3.

$x_3 = \dfrac{x_1 + x_2}{2} = \dfrac{3+2.5}{2} = 2.75$ and proceeding similarly we get x_4=2.875 and x_5=2.9375 and the process is continued for better accuracy.

Example 3.2: Solve $f(x) = x^3 - 2x^2 - 4$ by the method of interval halving procedure.

Solution: Given $f(x) = x^3 - 2x^2 - 4$, at first we have to find two numbers a = 2 and b = 3 such that f(a) = f(2) = –4 and f(b) = f(3) = 5 are of opposite signs.

Hence the root lies between a = 2 and b = 3. Then for iteration purpose, let us follows an alternate way of expressing the problem in tabular form as follows:

Iteration No	A	B	$x = \dfrac{a+b}{2}$	f(a)	f(b)	f(x)
1	2	3.00000	2.50000	-ve	+ve	-ve (1.8750
2	2.50000	3.00000	2.75000	-ve	+ve	+ve (1.6719)
3	2.50000	2.75000	2.62500	-ve	+ve	+ve (0.3066)
4	2.50000	2.62500	2.56250	-ve	+ve	-ve (0.3.640)
5	2.56250	2.62500	2.59375	-ve	+ve	-ve (0.0055)
6	2.59375	2.62500	2.60938	-ve	+ve	+ve (0.1488)
7	2.59375	2.60938	2.60157	-ve	+ve	+ve (0.0719)
8	2.59375	2.60157	2.59765	-ve	+ve	+ve (0.0329)
9	2.50375	2.59760	2.59570	-ve	+ve	+ve (0.0136)
10	2.50375	2.59570	2.59470	-ve	+ve	-ve (0.0040)

Hence, the root correct to three decimal places is x = 2.5947.

Example 3.3: Find a root of $f(x) = xe^x - 1 = 0$, using Bisection method, correct to three decimal places.

Solution: Let, $f(0) = 0.e^0 - 1 = -1 < 0$, $f(1) = 1.e^1 - 1 = 1.7183 > 0$, Hence a root of f(x)=0 lies in between 0 and 1. Therefore, $x_1 = \dfrac{0+1}{2} = 0.5$, $f(0.5)=0.5.e^{0.5}$ $-1 = -0.1756$, Hence the root lies in between 0.5 and 1.

$x_2 = \dfrac{0.5+1}{2} = 0.75$, proceeding like this, we get the sequence of approximations as follows.

Iteration No	a	B	$x = \dfrac{a+b}{2}$	f(a)	f(b)	f(x)
1	0	1	0.5	-ve	+ve	-ve (-1.1756)
2	0.5	1	0.75	-ve	+ve	+ve (0.58775)
3	0.5	0.75	0.625	-ve	+ve	+ve (0.3066)
4	0.5	0.625	0.5625	-ve	+ve	-ve (0.3.640)
5	0.625	0.5625	0.59375	-ve	+ve	-ve (0.0055)
6	0.59375	0.5625	0.5781	-ve	+ve	+ve (0.1488)
7	0.5625	0.5781	0.5703	-ve	+ve	+ve (0.0719)
8	0.5625	0.5703	0.5664	-ve	+ve	+ve (0.0329)
9	0.5703	0.5664	0.5684	-ve	+ve	+ve (0.0136)
10	0.5664	0.5684	0.5674	-ve	+ve	-ve (0.0040)
11	0.5664	0.5674	0.5669	-ve	+ve	-ve (-0.00067)
12	0.5674	0.5669	0.5672	-ve	+ve	+ve (0.00016)
13	0.5669	0.5672	0.5671,	-ve	+ve	-ve (-0.00012)

Hence, the required root correct to three decimal places is, x = 0.567.

3.4 Regula Falsi or False Position Method

This is one among the oldest method of computing the real root of a numerical equation f(x) = 0 and it is almost a replica of bisection method. Draw the graph of the curve y = f(x) through x = a and x = b such that f(a) and f(b) are of opposite signs. Then x = a, is the abscissa of the point C where the graph of y = f(x) meets the x-axis. Let the chord AB meets the x-axis at a point whose abscissa is x_1 which is taken as the first approximation of the root.

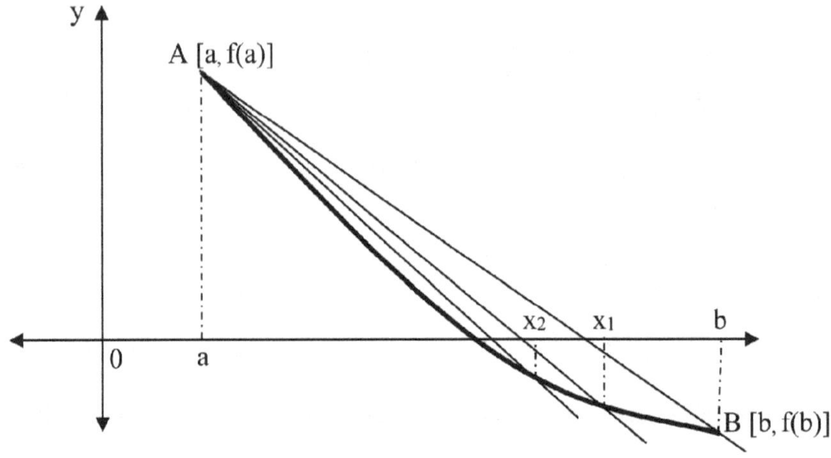

Fig. 3.2: Regula Falsi method

To obtain the value of x, consider the equation of the line AB (two point form) through

[a, f (a)] as bellow:

$$y - f(a) = m(x - a)$$

$$\Rightarrow y - f(a) = \frac{f(b) - f(a)}{b - a}(x - a) \text{ as } m = \frac{f(b) - f(a)}{b - a}$$

But it meets the x-axis at x = a where y = 0

$$0 - f(a) = \frac{f(b) - f(a)}{b - a}(x_1 - a)$$

$$\Rightarrow x_1 = a - \frac{b - a}{f(b) - f(a)} f(a)$$

Successive application of this process gives $x_1, x_2, \ldots\ldots\ldots\ldots, x_n$. Such points leading to an accurate value of x. This iteration process is known as the method of false position and its **rate if convergence equals 1.618,** which is faster than that of Bisection Method. Sometimes this method is also known as **Method of Variable Secant.**

General expression representing $(n + 1)^{th}$ iterated value $x_n + 1$ is as follows:

$$x_{n+1} = x_n - \frac{x_n - x_{n-1}}{f(x_n) - f(x_{n-1})} f(x_n)$$; provided that at each step $[f(x_n) - f(x_{n-1})] < 0$.

Example 3.4: Use Regula-Falsi Method to find a real root of the equation $x \log_{10}x - 1.2 = 0$

correct to five places of decimal.

Solution: Let $f(x) = x\log_{10}x - 1.2 = 0$

Now $f(2) = 2\log_{10}2 - 1.2 = 2 \times 0.30103 - 1.2 = -0.59794$ (–ve)

$f(3) = 3\log_{10}3 - 1.2 = 3 \times 0.47712 - 1.2 = 0.23136$ (+ve)

Thus, there is at least one root of the given equation which lies in the interval (2, 3).

Now $x_1 = a - \frac{b - a}{f(b) - f(a)} f(a)$ for a=2 and b=3

$$\Rightarrow x_1 = 2 - \frac{3 - 2}{0.23236 - (-0.59794)}(-0.59794) = 2.721$$

$\Rightarrow f(x_1) = f(2.721) = 2.721 \log_{10}2.721 - 1.2 = 2.721 \times 0.43472 - 1.2 = 0.0171$

Now, $f(x_1)$ is –ve and f (b) is +ve, so the root lies in between 2.721 and 3.

Here for the next iteration, a = 2.721 and b = 3.

$$x_2 = 2.721 - \frac{3 - 2.721}{0.23236 - (-0.0172)}(-0.0172) = 2.740201$$

Now $f(x_2) = f(2.740201) = -0.0004$

Clearly, the root lies between x(= a)=2.740201 and x(=b)=3.

$$x_3 = 2.74201 - \frac{3 - 2.740201}{0.23236 - (-0.0004)}(-0.0004) = 2.74066$$, which is (+ve)

Now, f(2.721201) is –ve and f(2.740201) is +ve, so we can proceed for further iterations for more accuracy. But, here we observe that the two consecutive values of x are same up to 3 decimal place and hence, 2.74065 is the correct values of x up to 5 decimal places.

Example 3.5: Using Regula Falsi Method, compute the real root of the equation $xe^x - 2 = 0$ correct up to three decimals places.

Solution: In $f(x) = xe^x - 2$; for $x = 1$, $f(1) = 1 \times e^1 - 2 = 0.718$ (+ve)

for $x = 0.5$, $f(0.5) = 0.5 \times e^5 - 2 = -1.17$ (–ve)

As the value of f(x) at x = 0.5 is –ve and at x = 1.0 it is +ve. Therefore, the root lies between x = 0.5 and x = 1.

Ist step: $x_0 = 0.5$ and $x_1 = 1$

$$\Rightarrow x_2 = x_0 - \frac{x_1 - x_0}{f(x_1) - f(x_0)} f(x_0)$$

$$\Rightarrow x_2 = 0.5 - \frac{1 - 0.5}{0.718 - (-1.17)}(-1.17) = 0.8098$$

$\Rightarrow f(x_2) = f(0.8098) = 0.8098\ e^{0.8098} - 2 = -0.18$

2nd step: Root lies between $x_1 = 1$ and $x_2 = 0.8098$

Therefore

$$x_3 = x_2 - \frac{x_1 - x_2}{f(x_1) - f(x_2)} f(x_2)$$

$$\Rightarrow x_3 = 0.8098 - \frac{1 - 0.8098}{0.718 - (-0.18)}(-0.18) = 0.8479$$

$\Rightarrow f(x_2) = f(0.8479) = 0.8479 - e^{0.8479} - 2 = -0.02$ (–ve)

3rd step: Next root lies between $x_3 = 0.847$ and $x_1 = 1$

$$x_4 = x_3 - \frac{x_1 - x_3}{f(x_1) - f(x_3)} f(x_3)$$

$$\Rightarrow x_4 = 0.8479 - \frac{1 - 0.8479}{0.718 - (-0.02)}(-0.02) = 0.8519$$

$\Rightarrow f(x_4) = 0.8519\ e^{0.8519} - 2 = -0.0031$

4th step: Next value of x in between $x_1 = 1$ and $x_4 = 0.8519$

$$x_5 = x_4 - \frac{x_1 - x_4}{f(x_1) - f(x_4)} f(x_4)$$

$$\Rightarrow x_5 = 0.8519 - \frac{1 - 0.8519}{0.718 - (-0.0031)}(-0.0031) = 0.8525$$

Approximate root is correct to 4 decimal of place, 0.8525.

Example 3.6: Find the equation $x^3 – 5x – 7 = 0$ which lies between 2 and 3 by the method of false position.

Solution: Let $f(x) = x^3 - 5x - 7$

Now by putting, $x = 2$ and $x = 3$, we get

$f(2) = (2)3 - 5(2) - 7 = -9$ and $f(3) = (3)3 - 5(3) - 7 = 5$

Hence the roots lies between 2 and 3, Let us take $x_0 = 2$ and $x_1 = 3$.

$$x_2 = x_0 - \frac{x_1 - x_0}{f(x_1) - f(x_0)} f(x_0)$$

$$\Rightarrow x_2 = 2 - \frac{2-3}{5-(-9)}(-9) = 2.64286$$

and $f(2.64286) = (2.64286)^3 - 5(2.64286) - 7 = 18.45960 - 13.2143 - 7 = -1.7547$

Hence, the root of equation lies between 2.64286 and 3.

For next iteration, $x_2 = 2.64286$ and $x_1 = 3$, so that

$$x_3 = x_1 - \frac{x_2 - x_1}{f(x_2) - f(x_1)} f(x_1)$$

$$\Rightarrow x_3 = 2.64286 - \frac{3 - 2.64286}{5-(-1.7547)}(-1.7547) = 2.73564$$

and $f(2.73564) = (2.73564)^3 - 5(2.73564) = -0.2054(-ve)$.

For next iteration, root lies between $x_3 = 2.73564$, $x_1 = 3$.

$$x_4 = x_2 - \frac{x_3 - x_2}{f(x_3) - f(x_2)} f(x_2)$$

$$\Rightarrow x_4 = 2.75364 - \frac{3 - 2.75364}{5-(-0.2054)}(-0.2054) = 2.7461$$

and $f(2.7461) = (2.7461)3 - 5(2.7461) - 7 = -0.0219$.

For next iteration, take $x_4 = 2.7461$, $x_1 = 3$

$$x_5 = 2.7461 - \frac{3 - 2.7461}{5-(-0.2019)}(-0.2019) = 2.7472$$

and $f(2.7472) = (2.7472)3 - 5(2.7472) - 7 = -0.00258$.

Hence again root lies between 2.7472 and 3.

$$x_6 = 2.7472 - \frac{3-2.7472}{5-(-0.00258)}(-0.00258) = 2.7473$$

Again putting x = 2.7473 in first equation, we get

f(2.7473) = (2.7473)3 – 5(2.7473) – 7 = –0.00082.

Hence, the required root up to 4 decimal places is = 2.7473.

3.5 Fixed Point Iteration Method

In Numerical analysis, it is a method of computing fixed points by doing number of iteration to the functions. In this method, at first, we have to convert your function f(x) into x = φ(x) form. Then iterate the value of x to the function φ(x) until you get the desire precision such that the precision value should be repeated after you again iterate it. We can make lots of functions in the form of x = φ(x) by using the f(x). Then, how can we say that every function will give the root of the f(x). So we have to need a test called "Condition for Convergence".

The condition of convergence is as follows: If α be the root of the f (x) which is equivalent to x = φ(x) and α is contained in the interval l of φ(x) and φ′(x) < 1 for all x belongs to l. Then, if x_0 is any point in l so that, the sequence defined by $x_n = \varphi(x_{n-1}) \mid n \geq 1$ will converges to the unique fixed point x in l . Thus, that unique fixed point x of φ(x) will be the root of f (x).

Example 3.7: Find the roots of the equation f (x) = $x^3 + x + 1$? Correct upto two decimal places.

Solution: Given, equation f(x)=x^3+x+1. To find the root of this equation, we say f(x)=0 i.e., $x^3 + x + 1 = 0$.

By testing on f(a), f(b) < 0. We estimated, the root should lies on [0; 1].

There are many ways to change the equation to the fixed point form x = φ(x) using simple algebraic manipulation.

$x = \varphi_1(x) = 1 - x^3$.

$x = \varphi_2(x) = (1 - x)^{1/3}$.

Taking the derivative of above forms with respect to x. We get $\varphi_1'(x) = -3x^2$.

and $\phi_2' = \dfrac{-1}{3(1-x)^{\frac{2}{3}}}$

Test on the value of x = 0:8. On putting in above equation and taking absolute value of it. We get

$$|\varphi_1^{/}(0.8)| = |-1.92| > 1$$

$$|\varphi_2^{/}(0.8)| = |-0.97| < 1$$

Since, >1, so we reject $\varphi_1(x)$ and we choose $\varphi_2(x)$ for iteration.

3.6 Iteration Method

Starting with an initial approximation to the solution of the problem, if the approximate solution is improved using some rule (formula) repeatedly then such a method is called an Iterative method.

Since the scheme is used repeatedly, there must be some convergence and stopping criteria attached with every iterative method. One of the convergence and stopping criteria is the maximum of the absolute value of the difference between two consecutive iterative values is less than a predefined tolerance limit. Mathematical, this is equivalent to the infinity norm of the difference between two consecutive iterative solutions is smaller than a predefined tolerance limit.

Any iterative method, to solve a linear system of equations, can be written as:

$x^{[k+1]} = G\, x^{[k]} + C$, where x is the unknown to be computed, G is the iteration matrix and C is a column vector, k is a iteration number varying from k = 0, 1, 2, 3,.....

The iterative matrix of the above iteration process is obtained from the coefficient matrix A by splitting it into various forms, For example consider the splitting,

A = D + L + U, where, D contains the diagonal elements of A, L contain the elements of A whose row number is greater than their column and U contains the elements of A whose row number is smaller than their column number

For example, if the matrix A is

$$A = \begin{pmatrix} 1 & 2 & 3 \\ 4 & 6 & 5 \\ 7 & -11 & 8 \end{pmatrix}$$

Then the decomposition discussed in the earlier slide is equivalent to

$$D = \begin{pmatrix} 1 & 0 & 0 \\ 0 & 6 & 0 \\ 0 & 0 & 8 \end{pmatrix}, L = \begin{pmatrix} 0 & 0 & 0 \\ 4 & 0 & 0 \\ 7 & -11 & 0 \end{pmatrix}, U = \begin{pmatrix} 0 & 2 & 3 \\ 0 & 0 & 5 \\ 0 & 0 & 0 \end{pmatrix}$$

Now, let us compute the iteration matrices for the standard iterative schemes

Jacobi Method:

Jacobi iterative scheme is generated by writing:

Ax=b or (D+U+L)x=b or Dx = -(U+L)x+b or x= $D^{-1}(U+L)+D^{-1}b$ or $x^{[k+1]}=G_{joc}x^{[k]}+C$, where, $G_{joc} = -D^{-1}(U+L)$ is the iteration matrix for the Jacobi method.

After choosing the initial approximation to the solution as x[0], the Jacobi method is implemented using

$$x_i^{[k+1]} = \frac{1}{a_{ii}}\left(b_i - \sum_{\substack{j=1 \\ j=i}}^{n} a_{ij}x_j^{[k]}\right), i = 1.2.3.....n \quad for\ k = 0,1,2,3,....$$

Example 3.8: Starting with $x^{[0]} = \begin{pmatrix} 0.5 \\ 0.5 \\ 0.5 \end{pmatrix}$ solve $\begin{pmatrix} 5 & 2 & 1 \\ 4 & 11 & 5 \\ 7 & 8 & 16 \end{pmatrix}\begin{pmatrix} x_1 \\ x_2 \\ x_3 \end{pmatrix} = \begin{pmatrix} 8 \\ 20 \\ 31 \end{pmatrix}$ using

Jacobi method (use $1.e^{-0.4}$ as the tolerance to stop the iterative process).

The Jacobi method generates the following sequence:

Iteration	X_1	X_2	X_3
0	0.5	0.5	0.5
1	1.3	1.1182	0.8097
2	0.9908	1.0899	0.9591
3	0.9722	1.0287	0.9978
5	0.9975	0.9998	1.0012
8	1.0000	1.0000	1.0000
9	1.0000	1.0000	1.0000

That is, Gauss Seidel method requires only 9 iterations to converge to the 10^{-4} accuracy.

3.7 Newton-Raphson Method

This method of successive approximation is due to English Mathematician and Physicist, Sir Issac Newton (1642-1727) is also called **fixed point method.** This technique is very useful for finding the root of the equation of the form f (x) = 0, where x is real and f (x) is an easily differential function.

To derive the formula for computing the root by this method, let x_0 denote the approximate value of the desired root and let h denote the correction which must be applied to x_0 to give the exact value of the root, so that $x = x_0 + h$.

The equation, f (x) = 0 then becomes f $(x_0 + h) = 0$, Expanding by Taylor's theorem, we get

$$0 = f(x_0 + h) = f(x_0) + hf'(x_0) + \frac{h^2}{2!} f''(x_0 + \theta h), 0 \le \theta \le 1$$

Now if h is relatively small, we may neglect the term containing h^2 and get the simple relation

$$f(x_0) + hf'(x_0) = 0 \text{ implying } h = \frac{f(x_0)}{f'(x_0)}$$

So that the first iterated value of x i.e., $x_1 = x_0 + h = x_0 - \frac{f(x_0)}{f'(x_0)}$

The next improved value roots is then $x_2 = x_1 - \frac{f(x_1)}{f'(x_1)}$

and so on, $x_{n+1} = x_n - \frac{f(x_n)}{f'(x_n)}$

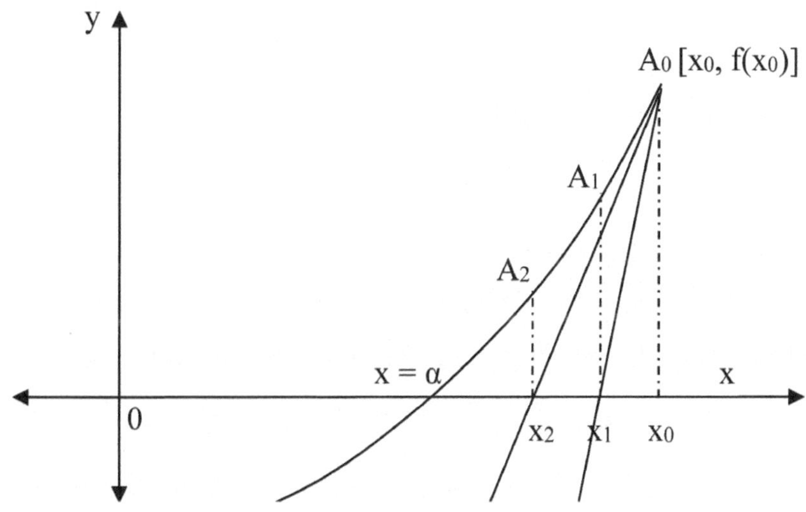

Fig. 3.3: Newton-Raphson method

Geometrical Significance: Let x_0 be a point near the exact α of the equation, f (x) = 0. Then in the given figure we can see that tangent at $A_0(x_0, f\,x_0)$ is y – $f(x_0) = f(x_0)(x - x_0)$. It cuts the x-axis at $x_1 = x_0 - \frac{f(x_0)}{f'(x_0)}$. (Since at $x = x_1$, y_1

= 0) which is the first approximation of the root α. If A_1 is the point corresponding to x_1 on the curve, on continuing, the tangent at A_1 will cut the x-axis at x_2 which is nearer to the root α, a second approximation to the root. This way, by repeating, we approaches the root quite rapidly. Clearly, here we replace the part of the curve between the point A_0 and the x-axis by means of the tangent at A_0.

Example 3.9: Compute the real root of $x\log_{10}x - 1.2 = 0$ upto four decimal places.

Solution: For given $f(x) = x \log_{10} x - 1.2$, let us find

$f(1) = 1 \log_{10} 1 - 1.2 = -1.2$

$f(2) = 2 \log_{10} 2 - 1.2 = -0.5979$

$f(3) = 3 \log_{10} 3 - 1.2 = 0.2313$

So the root lies between 2 and 3 since for x = 2, f(x) is negative where as for x = 3, f(x) is positive.

Take $x_0 = 3$, $f(x_0) = 0.2313$, and

$$f(x_0) = 1.\log_{10} x + x\frac{1}{x}\log_{10} e = \log_{10} 3 + \log_{10} e = 0.4771 + 0.14342 = 0.9113$$

By using $\log_a^x = \log_a^x . \log_e^x$

Now by Newton Raphson Method for $x_0 = 3$,

$$x_1 = x_0 - \frac{f(x_0)}{f'(x_0)} = 3 - \frac{0.2313}{0.9113} = 2.7462$$

For next iteration, let us find $f(x_1)$ at x1 = 2.7462 i.e

$f(x_1) = 2.7462 \times \log_{10}(2.7462) - 1.20 = 1.2048 - 1.20 = 0.0048.$

And $f'(x_1) = \log_{10}2.7462 + \log_{10}e = 0.4387 + 0.4342 = 0.8729$

$$x_2 = x_1 - \frac{f(x_1)}{f'(x_1)} = 2.7462 - \frac{0.0048}{0.8729} = 2.7407$$

So 2.7407 is the real root of the given equation correct upto four decimal places.

Example 3.10: Using Newton-Raphson Method, compute the real root of the following equation correct to four decimal places, $x = \sqrt{28}$.

Solution: Given => $(x^2 - 28) = 0$ => f(x) so that $f'(x) = 2x$. To start with, take x = 5, 6 and find $f(5) = 25 - 28 = -3$ and $f(6) = 36 - 28 = 8$. As value of f (x) at x = 5 is –ve, where as at x = 6 it is +ve, therefore, the root lies between 5 and 6, and near to 5.

Therefore for $x_0 = 5$, $f(x_0) = 3$, $f'(x_0) = 10$

$$x_1 = x_0 - \frac{f(x_0)}{f'(x_0)} = 5 - \frac{-3}{10} = 5.300$$

Next for second approximation,

$x_1 = 5.30$, $f(x_1) = x_2 - 28 = (5.30)^2 - 28 = 0.09$, $f'(x_1) = 2 \times 5.30 = 10.60$

$$x_2 = x_1 - \frac{f(x_1)}{f'(x_1)} = 5.300 - \frac{0.09}{10.60} = 5.2915$$

Now for third approximation ,

$x_2 = 5.0915$, $f(x_2) = (5.2915)^2 - 28 = -0.000027$, $f'(x_2) = 2 \times 5.2915 = 10.583$.

$x_3 = x_2 - \frac{f(x_2)}{f'(x_2)} = 5.219 - \frac{0.000027}{10.583} = 5.291502$, which is correct upto 4 decimal places.

3.7.1 Convergence of Newton-Raphson Method

We know the sequence of iterations $\{x_n\}_{x=1}$ generated by Newton's method can also be considered as iterations generated by Fixed Point iteration with function.

$$g_{newton}(x) = x - \frac{f(x)}{f'(x)}$$

By the fixed point theorem, the sequence of iterations $\{x_n\}_{x=1}$ converges to x^* if

$|g'(x)| \leq \kappa \leq 1$ for all x in [a, b]

where x_0 in [a, b] and the rate of convergence is $0(\kappa^n)$.

Let us now investigate the conditions on f(x) under which the property $|g'_{newton}(x)| < 1$ for $x \varepsilon [a, b]$ is satisfied.

i.e., $$g'_{newton}(x) = x - \frac{f'(x)f'(x) - f(x)f''(x)}{[f'(x)]^2}$$

$$g'_{newton}(x) = 1 - 1 + \frac{f(x)f''(x)}{[f'(x)]^2} = \frac{f(x)f''(x)}{[f'(x)]^2}$$

It is known that $f(x^*) = 0$. If $f'(x) \neq 0$, $|f'(x)| \leq \mu_1$ for all x in [a, b], then there exists an interval i.e., $\left[\bar{a},\bar{b}\right] \subseteq [a,b]$, such that $\left|\frac{f(x)f''(x)}{[f'(x)]^2}\right| \leq \kappa < 1$, and for x_0 is in $\left[\bar{a},\bar{b}\right]$, $\lim_{x\to\alpha} x_n = x^*$

3.7.1.1 Convergence property of Newton's method

Theorem: Let f, f' and f'' be continuous in [a, b]. If x^* is in [a, b] and $f'(x^*) \neq 0$, then there exist a $\delta > 0$ such that $\{x_n\}$ generated by Newton's method converges to x^* for any x_0 in $[x^*-\delta, x^*+\delta]$.

It is to be noted that

(a) The convergence of Newton's method on δ depends on $\lim_{x\to\infty} x_n = x^*$.

(b) By the fixed point iteration,

$x_n = x^* + O(K^n)$, where $\left|\frac{f(x)f''(x)}{[f'(x)]^2}\right| \leq K < 1$ for all x in $[x^*-\delta, x^*+\delta]$.

Example 3.11: Determine if the iterations generated by Newton's method for solving $x^2-2=0$ converges.

Solution: $f(x) = x^2-2$, $f'(x) = 2x$, $f''(x) = 2$.

$$|g'_{Newton}(x)| = \left|\frac{f(x)\,f''(x)}{[f'(x_n)]^2}\right| = \left|\frac{2(x^2-2)}{2x^2}\right| = \frac{1}{2}\left|1-\frac{2}{x^2}\right| \leq |g'_{Newton}(1)| = \frac{1}{2} = K$$

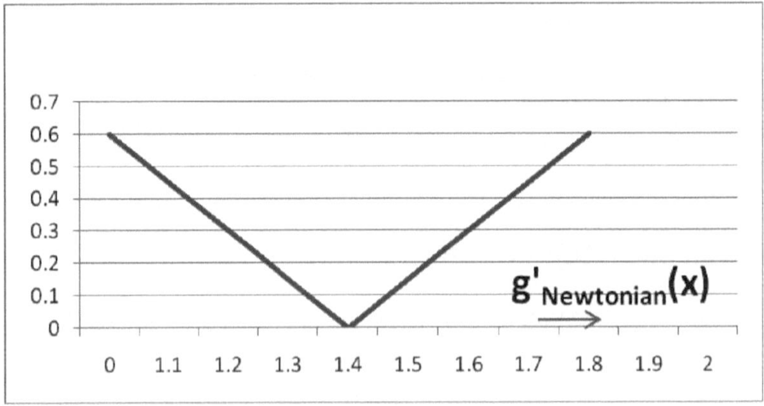

Fig. 3.4: Solution of example

The above graph concludes

(a) $\lim\limits_{x\to\infty} x_n = \sqrt{2}$ and

(b) $x_n = \sqrt{2} + O\{(0.5)^n\}$

3.7.1.2 Order of Convergence of Newton Raphson Method

Theorem 3.7.1.2.1: Modified Fixed Point Theorem: Let p be fixed point of g(x). If $g'(p) = 0$ and $g''(p) \neq 0$, then $p_n \to p$ quadratically ($\alpha = 2$) with an asymptotic error constant $\lambda = 0.5|g''(p)|$

Proof: Assuming that $g'(p) = 0$ but $g''(p) \neq 0$, so by the Taylor's theorem we have for $x \neq p$,

$$g(x) = g(p) + \frac{g'(p)}{1!}(x-p) + \frac{g''(c(x))}{2!}(x-p)^2,$$ where c(x) is between x and p.

Then for $x = p_n$, $$g(p_n) = g(p) + \frac{g'(p)}{1!}(p_n - p) + \frac{g''(c(x))}{2!}(p_n - p)^2$$

$$g(p_n) = g(p) + \frac{g''(c(p_n))}{2!}(p_n - p)^2,$$ where $c(p_n-1)$ is between x and p_{n-1}.

$$|p_n - p| = |g(p_{n-1}) - g(p)| = \left|g(p) + \frac{g''(c(p_{n-1}))}{2!}(p_n - p)^2 - g(p)\right|$$

$$= \frac{1}{2}|g''(c(p_{n-1}))||p_n - p|^2$$

and $$\lim_{n\to\infty}\frac{|p_{n+1} - p|}{|p_n - p|^2} = \lim_{n\to\infty}\frac{1}{2}|g''(c(p_n))| = \frac{1}{2}\left|g''\left[\lim_{n\to\infty}(c(p_n))\right]\right| = \frac{1}{2}|g''(p)|$$

So the order of convergence is 2 and the asymptotic error constant is $\frac{1}{2}|g''(p)|$

Theorem 3.7.1.2.2: Let $\{x_n\}$ be generated by the Newton's method. If x^8 is a simple zero of f(x), then $x_n \to x^*$quadratically with an asymptotic error constant$\left|\frac{f''(x^*)}{2f'(x^*)}\right|$

Proof: By the modified fixed point theorem, we have $g''(x) = \frac{f(x)f''(x)}{[f'(x)]^2}$ then

$$g''(x) = \frac{[f'(x)f''(x) + f(x)f'''(x)][f'(x)]^2 - f(x)f''(x)[f'(x)f''(x)]}{[f'(x)]^2}$$

$$= \frac{[f'(x)f''(x) + f(x)f'''(x)][f'(x)] - 2f(x)[f''(x)]^2}{[f'(x)]^3}$$

$$g''(x^*)\,f(x^*) = 0 = \frac{[f'(x^*)]^2 f''(x^*)}{[f'(x^*)]^2} = \frac{f''(x^*)}{f'(x^*)}$$

Then the asymptotic error constant is $\frac{1}{2}|g''(x^*)| = \left|\frac{f''(x^*)}{2f'(x^*)}\right|$

3.7.2 Rate of Convergence of Newton Raphson Method

Suppose α represents the exact value of the root of $f(x) = 0$ and let x_i, x_{i+1} are the two successive approximations to the actual root α. If ε_i and ε_{i+1} are the corresponding error, then

$x_i = \alpha + \varepsilon_i$ and $x_{i+1} = \alpha + \varepsilon_{i+1}$

Then through Newton-Raphson iterative method

$$\alpha + \varepsilon_{i+1} = \alpha + \varepsilon_i - \frac{f(\alpha + \varepsilon_i)}{f'(\alpha + \varepsilon_i)}$$

$$\varepsilon_{i+1} = \varepsilon_i - \frac{f(\alpha + \varepsilon_i)}{f'(\alpha + \varepsilon_i)}$$

$$\varepsilon_{i+1} = \varepsilon_i - \frac{f(\alpha) + \varepsilon_i f'(\alpha) + \left(\frac{\varepsilon_i^{\,2}}{2}\right) f''(\alpha) + \ldots\ldots\ldots}{f'(\alpha) + \varepsilon_i f''(\alpha) + \ldots\ldots\ldots}$$

$$= \varepsilon_i - \frac{\varepsilon_i f'(\alpha) + \frac{\varepsilon_i^{\,2}}{2} f''(\alpha) + \ldots\ldots\ldots}{f'(\alpha) + \varepsilon_i f''(\alpha) + \ldots\ldots\ldots}$$

$$= \frac{1}{2}\left[\frac{\varepsilon_i^{\,2} f''(\alpha)}{f'(\alpha) + \varepsilon_i f''(\alpha) + \ldots\ldots\ldots}\right]$$

$$= \frac{1}{2}\left[\frac{\varepsilon_i^{\,2} f''(\alpha)}{f'(\alpha)\left(1 + \varepsilon_i \frac{f''(\alpha)}{f'(\alpha}\right) + \ldots\ldots\ldots}\right] = \varepsilon_{i+1} \cong \frac{f''(\alpha)}{2f'(\alpha)}$$

Hence this equation represents that the error at each stage is proportional to the sequence of the error in the previous stage, therefore the convergence of Newton-Raphson method is quadratic.

3.7.3 Modified Newton-Raphson Method

Let x^* be a zero of multiplicity 2 of on [a, b]. Then we know $f'(x^*) = 0$, though $f(x_n) \neq 0$, as $x_n \to x^*$, $f'(x_n) \to f'(x^*) = 0$, so $f'(x_n) \approx 0$ for large n. In this case, does $\{x_n\}$ still converge to x^*.

Let $g(x) = x - \frac{f(x)}{f'(x)}$

For convergence, we need to find out

$$\lim_{x \to x^*} |g'(x)| < 1$$

$$g(x) = 1 - \frac{[f'(x)]^2 - f(x)f''(x)}{[f'(x)]^2} = \frac{f(x)f''(x)}{[f'(x)]^2}$$

$$\lim_{x\to x^*} g'(x) = \lim_{x\to x^*}\frac{f(x)f''(x)}{[f'(x)]^2}$$

$$= \lim_{x\to x^*}\frac{f'(x)f''(x)+f(x)f'''(x)}{2f'(x)f''(x)}$$

$$= \frac{1}{2}+\lim_{x\to x^*}\frac{f(x)f'''(x)}{2f'(x)f''(x)}$$

$$= \frac{1}{2}+\frac{1}{2}\lim_{x\to x^*}\frac{f'(x)f'''(x)+f(x)f^4(x)}{[f''(x)]^2+f'(x)f'''(x)}$$

$$= \frac{1}{2} < 1$$

So, $\{x_n\}$ still converges to x^*, however since value is not equal to 0, $x_n \to x^*$ linearly instead of quadratically. If $f'(x^*) = 0$, $f''(x^*) = 0$ but $f'''(x^*) \neq 0$. Then

$$\lim_{x\to x^*} g'(x) = \frac{1}{2}+\frac{1}{2}\lim_{x\to x^*}\frac{f'(x)f'''(x)+f(x)f^4(x)}{[f''(x)]^2+f'(x)f'''(x)}$$

$$= \frac{1}{2}+\lim_{x\to x^*}\frac{f''(x)f'''(x)+2f'(x)f^4(x)+f(x)f^5(x)}{2[3f''(x)f'''(x)+f'(x)f^4(x)]}$$

$$= \frac{1}{2}+\frac{1}{2}\lim_{x\to x^*}\frac{[f'''(x)]^2+3f''(x)f^4(x)+3f'(x)f^5(x)+f'(x)f^6(x)}{[3\{f'''(x)\}^2+4f''(x)f^4(x)+f'(x)f^5(x)]}$$

$$= \frac{1}{2}+\frac{1}{2(3)} = \frac{2}{3}$$

If $f'(x^*) = 0$, $f''(x^*) = 0$, $f'''(x^*) = 0$, but $f^4(x) \neq 0$, then

$$\lim_{x\to x^*} g'(x) = \frac{1}{2}+\frac{1}{2}\lim_{x\to x^*}\frac{[f'''(x)]^2+3f''(x)f^4(x)+3f'(x)f^5(x)+f'(x)f^6(x)}{[3\{f'''(x)\}^2+4f''(x)f^4(x)+f'(x)f^5(x)]}$$

$$= \frac{1}{2}+\frac{1}{2}\lim_{x\to x^*}\frac{\begin{bmatrix}2f'''(x)f^4(x)+3f'''(x)f^4(x)+3f''(x)f^5(x)+f'''(x)f^6(x)+\\ 3f''(x)f^5(x)+3f'(x)f^6(x)+f'(x)f^7(x)\end{bmatrix}}{\left[6f'''(x)f^4(x)+4f'''(x)f^4(x)+f''(x)f^5(x)+4f''(x)f^5(x)+f'(x)f^6(x)\right]}$$

$$= \frac{1}{2}+\frac{1}{2}\lim_{x\to x^*}\frac{\left[5f'''(x)f^4(x)+6f''(x)f^5(x)+f'''(x)f^6(x)+3f'(x)f^6(x)+f'(x)f^7(x)\right]}{\left[10f'''(x)f^4(x)+5f''(x)f^5(x)+f(x)f^6(x)\right]}$$

$$= \frac{1}{2}+\frac{1}{2}\lim_{x\to x^*}\frac{5\left[f^4(x)\right]^2+5f'''(x)f^5(x)+\ldots\ldots\ldots\ldots}{10\left[f^4(x)\right]^2+10f'''(x)f^5(x)+\ldots\ldots\ldots\ldots} = \frac{3}{4}$$

Corollary: If $f'(x^*) = 0$, $f''(x^*) = 0$, ..., $f^m(x) = 0$, but $f^{(m+n)}(x^*) \neq 0$, then

$$\lim_{x \to x^*} g'(x) = \frac{m+1}{m}$$

3.7.4 Rate of Convergence of Modified Newton-Raphson Method

Suppose α is the root of $f(x) = 0$ equation. Consider, $\varepsilon_i = x_i - \alpha$.

Hence $x_{i+1} = \varphi(x_i) = \varphi(\varepsilon_i + \alpha)$, or

$$\varepsilon_{i+1} + \alpha = \phi(\alpha) + \varepsilon_i \phi'(\alpha) + \frac{\varepsilon_i^{\,2}}{2!}\phi'(\alpha) + \frac{\varepsilon_i^{\,3}}{3!}\phi'''(\alpha) + ...$$

$$\text{Or } \varepsilon_{i+1} = \frac{\varepsilon_i^{\,3}}{3!}\phi'''(\alpha) + O(\varepsilon_i^{\,4})$$

This equation represents the rate of convergence of the modified Newton-Raphson method.

The above equation becomes; $\varepsilon_{i+1} = A\,\varepsilon_i^3$, where $A = \frac{1}{3!}\phi'''(\alpha)$

Example 3.12: Through modified Newton-Raphson method, find the real root near 2 of equation $x^4 - 11x + 8 = 0$ upto 5 decimal places.

Solution: $f(x) = x^4 - 11x + 8$

$f'(x) = 4x^3 - 11$

$f''(x) = 12x^2$

At $x_0 = 2$,

$f(x_0) = (2)^4 - 11(2) + 8 = 2$

$f'(x_0) = 4(2)^3 - 11 = 21$

$f''(x_0) = 12(2)^2 = 48$

The modified Newton-Raphson method is $x_{n+1} = x_n - \frac{f(x_n)f'(x_n)}{[f'(x_n)]^2 - f(x_n)f''(x_n)}$

N	x_i	$f(x_i)$	$f'(x_i)$	$f''(x_i)$	x_{i+1}
0	2	2	21	48	1.87826
1	1.87826	–0.21505	15.50499	42.33437	1.89162
2	1.89162	–0.00405	16.07476	42.93891	1.89187
3	1.89187	$-1.4*10^{-6}$	16.08557	42.95034	1.89187

Therefore the root is 1.89187 of equation $f(x) = x^4 - 11x + 8$

3.8 Successive Approximation Method

Suppose equation f(x) = 0, whose roots are to be determined and thus the equation can be written as x= f(x). Suppose $x = x_0$ be an initial approximation to the desired root β. Then, the first approximation x_1 is; $x_1 = \phi(x_0)$. The second approximation is $x_2 = \phi(x_1)$; and successive approximations are $x_3 = \phi(x_2)$, $x_4 = \phi(x_3)$, ... , $x_n = \phi(x_{n-1})$. The sequence of approximations of x_1, x_2, x_3, ..., x_n always converge to the root of $x = \phi(x)$ and it can be shown that if $|\phi'(x)| < 1$, when x is sufficiently close to the exact value c of the root and $x_n \to c$ as $n \to \infty$. The convergence of $x_{i+1} = \phi(x_n)$, for $|\phi'(x)| < 1$.

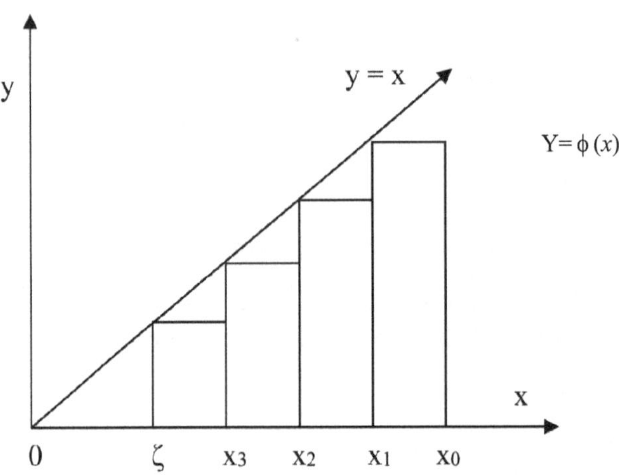

Fig. 3.5: Successive approximation method

Theorem: This theorem presents the convergence criteria for the iterative sequence of solution for the method. Consider α be the root of f(x) = 0, equivalent to $x = \phi(x)$, $\phi(x)$ is continuously differentiable function in an interval I containing the root x = α, if $|\phi'(x)| < 1$, then sequence of approximations x_0, x_1, x_2, x_3, ..., x_n will converge to the root α provided the inital approximation $x_0 \varepsilon$ 1.

Let α is the initial root of $x = \phi(x)$, then $\alpha = \phi(a)$ (1)

Suppose $x = x_0$ is the initial approximation, then $x_1 = \phi(x_0)$ (2)

From the equations (1) and (2), yield, $\alpha - x_1 = \phi(\alpha) - \phi(x_0)$ (3)

Through Lagrange's mean value theorem, equation (3) becomes:

$\alpha - x_1 = (\alpha - x_0)\,\phi'(\mu_0)$, for $x_0 < x_0 < \alpha$

Similarly the next approximations are

$\alpha - x_2 = (\alpha - x_1)\ \phi'(\mu_1)$, for $x_1 < \mu_1 < \alpha$

$\alpha - x_3 = (\alpha - x_2)\ \phi'(\mu_2)$, for $x_2 < \mu_2 < \alpha$

$\alpha - x_4 = (\alpha - x_3)\ \phi'(\mu_3)$, for $x_3 < \mu_3 < \alpha$

....

$\alpha - x_n = (\alpha - x_{n-1})\ \phi'(\mu_{n-1})$, for $x_{n-1} < \mu_{n-1} < \alpha$

On multiplying all the above equations, yielding:

$\alpha - x_n = (\alpha - x_0)\ \phi'(\mu_0)\ \phi'(\mu_1)\ \phi'(\mu_2)\ \phi'(\mu_3)\ \ldots\ldots\ldots\ \phi'(\mu_{n-1})$

If $\phi'(x_i) \leq K \leq 1$, for all I, then the above equation becomes

$|x_n - \alpha| \leq |x_0 - \alpha|\mu_0\ |\phi'(\mu_0)|\ \phi'(\mu_1)\ \ldots\ldots\ldots\ \phi'(\mu_{n-1}) \leq K, K\ \ldots\ldots\ K\ |x_0 - \alpha| \leq K^n$
$|x_0 - \alpha|$

As $K < 1$, so $K^n \to 0$, as $n \to \infty$, therefore $x_n \to \alpha$, for $x_0 \varepsilon I$.

3.8.1 Error Estimate in the Successive Approximation Method

Suppose $\varepsilon_n = x_n - \xi$, the error estimate at the nth iteration, then $\lim\limits_{n\to\infty}\left(\frac{\varepsilon_{n+1}}{\varepsilon_n}\right) = \varphi'(\varepsilon)$ is satisfied. It is known from mean value theorem that $|x_{n+1} - \xi| = |\varphi(x_n) - \varphi(\xi)| = |\varphi'(\xi_n)|\ |x_n - \xi|$, $\xi_n \in (x_n, \xi)$. So, $\varepsilon_{n+1} = \varepsilon_n$ $|\varphi'(\xi_n)|$, therefore $\lim\limits_{n\to\infty}\left(\frac{\varepsilon_{n+1}}{\varepsilon_n}\right) = \varphi'(\varepsilon)$

Therefore, the order of convergence is linear. But if $\varphi'(\xi) = 0$ and $\varphi''(\zeta) \neq 0$, then through Taylor's series expansion of ϕ in the neighbourhood of ξ is given by

$$\varphi_n = \varphi(\zeta) + (x_n - \zeta)\varphi'(\zeta) + \frac{(x_n - \zeta)^2}{\angle 2}\varphi''(\zeta) + \ldots\ldots\ldots$$

Which shows that $\varepsilon_{n+1} = \varepsilon_n \varphi'(\zeta) - \frac{\varepsilon_n^{\ 2}}{2}\varphi''(\zeta) + \frac{\varepsilon_n^{\ 3}}{6}\varphi''(\zeta) - \ldots\ldots\ldots$

And on using $x_{n+1} = \varphi(x_n)$ and $\varepsilon_n = |x_{n+1} - \xi|$, hence $\varepsilon_{n+1} = -\frac{\varepsilon_n^{\ 2}}{2}\varphi''(\zeta)$, on neglecting the terms with higher powers of ε_n, called quadratic convergence.

Example 3.13: Find the real roots of $x^3 - 2x - 3 = 0$, upto 3 decimal places through successive approximation method.

Solution: $f(x) = x^3 - 2x - 3 = 0$

At $x = 1$, $f(1) = 1^3 - 2(1) - 3 = -4 < 0$

At $x = 2$, $f(2) = 2^3 - 2(2) - 3 = 1 > 0$

Hence the root lies in between 1 and 2, as $f(1) < f(2)$, hence the initial approximation is $x_0 = 1$, yielding the equation; $x^3 = 2x + 3$, or $x = (2x + 3)^{1/2} = \varphi(x)$

The successive approximations of the root are

$x_1 = \varphi(x_0) = (2x_0 + 3)^{1/2} = \{2\ (1) + 3\}^{1/2} = 1.25992$

$x_2 = \varphi(x_1) = (2x_1 + 3)^{1/2} = \{2\ (1.25992) + 3\}^{1/2} = 1.31229$

$x_3 = \varphi(x_2) = (2x_2 + 3)^{1/2} = \{2\ (1.31229) + 3\}^{1/2} = 1.32235$

$x_4 = \varphi(x_3) = (2x_3 + 3)^{1/2} = \{2\ (1.32235) + 3\}^{1/2} = 1.32427$

$x_5 = \varphi(x_4) = (2x_4 + 3)^{1/2} = \{2\ (1.32427) + 3\}^{1/2} = 1.32463$

Hence the root (real) of $f(x) = 0$ is 1.324.

3.9 Secant Method

This method is similar, but an improvement over Newton-Raphson method. The main drawback of Newton-Raphson method is that it requires the determination of the derivatives of the function at several points and the calculations of these derivatives takes enough time, and in some cases the derivative may difficult to obtain and may not be available. But in secant method, the derivative of the function is approximated through finite differences as, rather than analytical calculations,

$$f'(x_n) = \frac{f(x_n) - f(x_{n-1})}{x_n - x_{n-1}}$$

where x_n and x_{n-1} are the two approximations to the root and not require the $f(x_n)\,f(x_{n-1}) < 0$ condition. Secant method is generated through Newton-Raphson method as

$$x_{n+1} = x_n - \frac{f(x_n)}{f'(x_n)}$$

$$x_{n+1} = x_n - \frac{x_n - x_{n-1}}{f(x_n) - f(x_{n-1})} f(x_n), n \geq 1$$

In this method, there are 2 approximations i.e., x_0 and x_1 for the root. This method is shown graphically where secant is drawn connecting $f(x_0)$ and $f(x_1)$. The point where it intersects the x-axis is x_{n+1} i.e., x_2. After again drawing secant connecting $f(x_1)$ and $f(x_2)$ in view to obtain x_{n+2} i.e., x_3. This process continuous until value remains same.

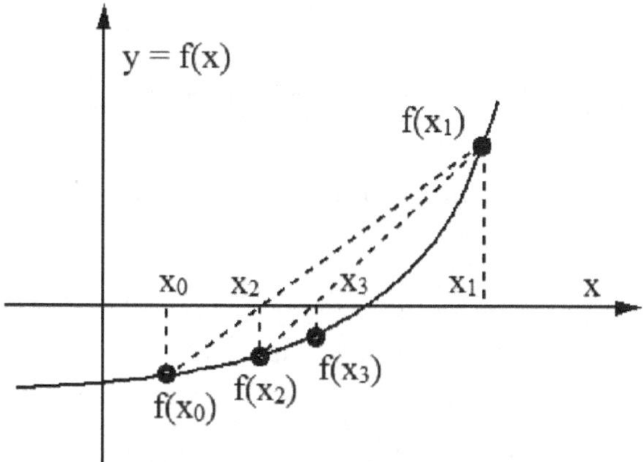

Fig. 3.6: Graphical representation of secant method

Drawback: This method fails if at any iteration, $f(x_n) = f(x_{n-1})$ and represents that it does not converges necessarily.

Example 3.14: Through secant method, find a root of the equation $x^3 - 2x - 5 = 0$.

Solution: The given equation is, $f(x) = x^3 - 2x - 5 = 0$

At $x = 2$, $f(2) = (2)^3 - 2(2) - 5 = -1$

At $x = 3$, $f(3) = (3)^3 - 2(3) - 5 = 16$

Hence the root lies in between 2 and 3. So the initial approximations are, $x_0 = 2$ and $x_1 = 3$.

$$x_2 = x_1 - \frac{x_1 - x_0}{f(x_1) - f(x_0)} f(x_1) = 3 - \frac{3-2}{16+1} 16 = 2.058823$$

At $x_2 = 2.058823$, $f(x_2) = -0.390799$

$$x_3 = x_2 - \frac{x_2 - x_1}{f(x_2) - f(x_1)} f(x_2) = 2.058823 - \frac{2.058823 - 3}{-0.390799 + 16}(-0.390799) = 2.081263$$

At $x_3 = 2.081263$, $f(x_3) = -0.147204$

$$x_4 = x_3 - \frac{x_3 - x_2}{f(x_3) - f(x_2)} f(x_3) = 2.094824$$

At $x_4 = 2.094824$, $f(x_4) = 0.003042$

$$x_5 = x_4 - \frac{x_4 - x_3}{f(x_4) - f(x_3)} f(x_4) = 2.094549$$

Hence the root is x = 2.094 upto three decimal places.

3.9.1 Convergence of the Secant Method

The secant method is, $x_{n+1} = x_n - \dfrac{x_n - x_{n-1}}{f(x_n) - f(x_{n-1})} f(x_n), n \geq 1$ (1)

Suppose δ is the exact root of equation f(x) = 0 and f(δ) = 0. The error at the nth iteration is,

$E = x_n - \delta$ (2)

Put the value of x_n in equation (1), resulting

$$E_{n+1} = E_n - \frac{(E_n - E_{n-1})}{f(E_n + \delta) - f(E_{n-1} + \delta)} f(E_n + \delta), n \geq 1 \quad (3)$$

$$E_{n+1} = E_n - \frac{(E_n - E_{n-1})[f(\delta) + E_n f''(\delta) + \left(\frac{E_n^2}{2}\right) f''(\delta) + ...}{(E_n - E_{n-1}) f''(\delta) + \frac{1}{2}(E_n^2 - E_{n-1}^2) f''(\delta) + ...}$$

$$E_{n+1} = E_n - \left[E_n + \frac{E_n^2}{2}\frac{f''(\delta)}{f''(\delta)} + ...\right]\left[1 + \frac{1}{2}(E_n + E_{n-1})\frac{f''(\delta)}{f''(\delta)} + ...\right]^{-1}$$

$$E_{n+1} = \frac{1}{2} E_n E_{n-1} \frac{f''(\delta)}{f'(\delta)} + O(E_n^2 E_{n-1} - E_n E_{n-1}^2) \quad (4)$$

$$E_{n+1} = c\, E_n\, E_{n-1} \quad (5)$$

where $c = \dfrac{1}{2}\dfrac{f''(\delta)}{f'(\delta)}$, is a non-linear equation, solved through $E_{n+1} = A E_n^p$ or

$E_{n+1} = A E_{n-1}^p$, resulting in $E_{n-1} = A^{-1/p} E_n^{1/p}$

Hence or $A E_n^p = c E_n E_n^{1/p} A^{-1/p}$ or $E_n^p = c E_n^{1+1/p} A^{-(1+1/p)}$ (6)

On equating the powers of E_n, resulting in, $p = 1 + \dfrac{1}{p}$ or $p = \dfrac{1}{2}(1 + \sqrt{5})$

When p is +ve, then p = 1.618 and $E_{n+1} = A\, E_n^{1.618}$

Hence the rate of convergence is 1.1618, which is lesser than Newton-Raphson method, hence is more efficient and time saving method as compared to Newton-Raphson method.

3.10 Muller's Method

This method is an iterative root finding method for solving system of linear equations of f(x) = 0 form. In secant method, roots are obtained by projecting a straight line to the x-axis through two function values. Muller's method is based on secant method, but uses three function points in view to construct a parabola and takes the intersection of the x-axis through 3rd function value with the parabola to the next approximation. This method is named after 'David Muller' in 1956.

The three initial points are (x_{n-2}, f_{n-2}), (x_{n-1}, f_{n-1}) and (x_n, f_n). A parabola is constructed that passes through these points and then the quadratic formula is used in view to find the point, where the parabola intercepts the x-axis, the root estimate.

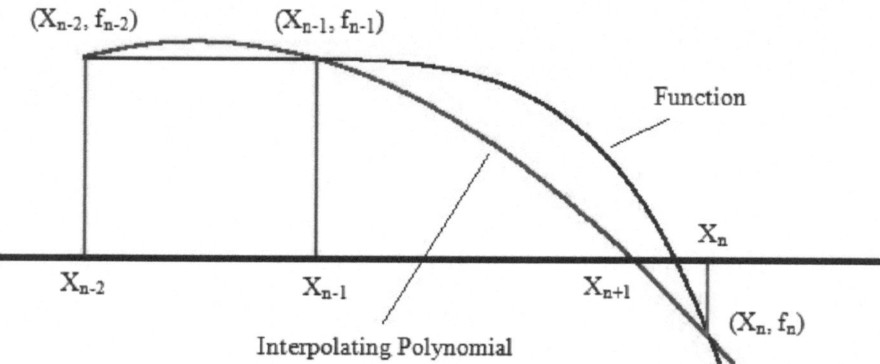

Fig. 3.7: Muller's method

Let n = 2, then the initial points are x_0, x_1 and x_2. Then determine the next approximation x_3 by considering the intersection at x-axis passes through (x_0, f_0), (x_1, f_1) and (x_2, f_2).

Suppose the quadratic polynomial is: $f(x) = a(x - x_2)^2 + b\,(x - x_2) + c$ (1)

The coefficients of equation (1) can be evaluated by substituting each of three points to give equations

$$f(x_0) = f_0 = a(x_0 - x_2)^2 + b\,(x_0 - x_2) + c \quad (2)$$

$$f(x_1) = f_1 = a(x_1 - x_2)^2 + b\,(x_1 - x_2) + c \quad (3)$$

$$f(x_2) = f_2 = a(x_2 - x_2)^2 + b\,(x_2 - x_2) + c \quad (4)$$

There are 3 equations [(2), (3) and (4)] and 3 variables (a, b and c). From equation (4), . Put the value of c in equation (2) and (3) to yield two equations with 2 unknown variables.

$$f_0 - f_2 = a(x_0 - x_2)^2 + b\,(x_0 - x_2) \tag{5}$$

$$f_1 - f_2 = a(x_1 - x_2)^2 + b\,(x_1 - x_2) \tag{6}$$

Algebraic manipulation can be used to solve the remaining coefficients a and b. Number of differences can be used as

$h_0 = x_1 - x_0$ and $h_1 = x_2 - x_1$

$$\delta_0 = \frac{f_1 - f_0}{x_1 - x_0} \text{ and } \delta_1 = \frac{f_2 - f_1}{x_2 - x_1}$$

Put these values in equations (5) and (6) yields

$$(h_0 + h_1)\,b - (h_0 + h_1)^2\,a = h_0\,\delta_0 + h_1\,\delta_1 \tag{7}$$

$$h_1\,b - h_1^2\,a = h_1\,\delta_1 \tag{8}$$

Hence

$$a = \frac{\delta_1 - \delta_0}{h_1 + h_0},$$

$b = a\,h_1 + \delta_1$ and

$c = f_2$

Hence x_3 can be determined through the use of formula which is used for calculation of parabolic roots, as

$$x_3 - x_2 = \frac{2c}{b \pm \sqrt{b^2 - 4ac}} \tag{9}$$

or $x_3 = x_2 + \dfrac{2c}{b \pm \sqrt{b^2 - 4ac}}$

The advantage of this usage of this formula is both real and complex roots can be located. In addition to this the error can be calculated as

$$\varepsilon_a = \frac{x_3 - x_2}{x_3} X 100$$

Equation (9) yields two roots, corresponding to the $\pm$ term in the denominator. The sign is chosen to agree with the sign of b. This choice will result in the largest denominator and hence will give the root estimate that is closest to x_2.

Once x_3 is determined, the process is repeated. This brings up the issue of which point is discarded. Two general strategies are used i.e.,

(a) If only real roots are being located, then choose the two original points that are nearest the new root estimate x_3.

(b) If both real and complex roots are being evaluated, a sequential approach is employed. Similar to secant method, x_1, x_2 and x_3 take place of x_0, x_1 and x_2.

The order of convergence is ~ 1.84 as compared to secant method (1.62 order of convergence) and Newton-Raphson method (2 order of convergence). So the speed of Muller's method is more than secant method, but less than Newton-Raphson method.

Example 3.15: Find a root of a equation $x^3 - 13x - 12$ using Muller's method, where the root lies in x_0, x_1 and x_2 = 4.5, 5.5 and 5 respectively.

Solution: The given function is $f(x) = x^3 - 13x - 12$

For $x_0 = 4.5$, $f(x_0) = f_0 = 20.625$

For $x_1 = 5.5$, $f(x_1) = f_1 = 82.875$

For $x_2 = 5$, $f(x_2) = f_2 = 48$

Number of differences; $h_0 = x_1 - x_0 = 5.5 - 4.5 = 1$, and $h_1 = x_2 - x_1 = 5 - 5.5 = -0.5$

$$\delta_0 = \frac{f_1 - f_0}{x_1 - x_0} = \frac{82.875 - 20.625}{5.5 - 4.5} = 62.25 \text{ and } \delta_1 = \frac{f_2 - f_1}{x_2 - x_1} = \frac{48 - 82.875}{5 - 5.5} = 69.75$$

The values of coefficients by putting the above values are

$$a = \frac{\delta_1 - \delta_0}{h_1 + h_0} = \frac{69.75 - 62.25}{-0.5 + 1} = 15,$$

$b = a h_1 + \delta_1 = 15(-0.5) + 69.75 = 62.25$

$c = f_2 = 48$

The square root of the discriminanr can be evaluated as

$$\sqrt{b^2 - 4ac} = \sqrt{(62.25)^2 - 4(15)(48)} = 31.54461$$

Then, because $|62.25 + 31.54461| > |62.25 - 31.54461|$, a positive sign is employed in the denominator and the new root is

$$x_3 = x_2 + \frac{2c}{b + \sqrt{b^2 - 4ac}} = 5 + \frac{-2(48)}{62.25 + 31.54461} = 3.976487$$

The error estimate is,

$$x_3 = x_2 + \frac{2c}{b+\sqrt{b^2-4ac}} = 5 + \frac{-2(48)}{62.25+31.54461} = 3.976487$$

This error is large, so new guesses are assigned, x_0 is replaced by x_1, x_1 is replaced by x_2 and x_2 is replaced by x_3. So the new iterations are; $x_1 = 5.5$, $x_2 = 5$ and $x_3 = 3.946487$, and the calculation is repeated. The results are

I	x_r	Error percentage
0	5	
1	3.946487	25.74
2	4.00105	0.6139
3	4	0.0262
4	4	0.0000119

The results shows that this method converges rapidly on the root $x_t = 4$.

3.11 Chebyshev Method

This method is an iterative method, is used for determination of simple roots of system of linear equations. This method is based on inverse interpolation so avoids computation of inner products/vectors as compared to other non-stationary methods. This method is named after Russian Mathematician 'Pafnuty Chebyshev'.

Suppose the equation, f(x) = 0, which can be solved, where f is a sufficiently smooth function. On expansion of equation through Taylor's series yields

$$f(x) = 0$$

$$f(x_n) + (x - x_n) f'(x_n) + \ldots\ldots = 0 \tag{1}$$

Equation (1) yields,

$$x = x_n - \frac{f(x_n)}{f'(x_n)} \tag{2}$$

Through equation (2), the $(n+1)^{th}$ approximation of the root is

$$x_{n+1} = x_n - \frac{f(x_n)}{f'(x_n)} \tag{3}$$

On expansion of equation (3) through Taylor's series upto 2nd order yields

$$f(x_n) + (x - x_n) f'(x_n) + \frac{(x - x_n)^2}{2} f''(x_n) = 0 \tag{4}$$

Through equation (4), the $(n+1)^{th}$ approximation of the root is

$f(x_{n+1}) = f(x_n) + (x_{n+1} - x_n) f'(x_n) + f''(x_n) = 0$ .... (5)

From equation (3), put the value of $(x_{n+1} - x_n)$ to the last term of equation (5) yields **Chebyshev's formula** (or **extended form of Newton-Raphson formula**) and corresponding the rate of convergence is cubic.

$$f(x_n) + (x_{n+1} - x_n) f'(x_n) + \frac{(x_{n+1} - x_n)^2}{2} f''(x_n) = 0 \tag{6}$$

3.12 Aitken's Δ^2 Method

Aitken Exploration or Aitken's Δ^2 method is a series acceleration method which is used for accelerating the rate of convergence of a sequence; useful most for sequence which is converging linearly. This method is named after 'Alexander Aitken' in 1926.

Consider equation whose roots are to be determined; $f(x) = 0$ (1)

Let x_{i-1}, x_i, x_{i+1} are the three successive approximations to the desired root α of the above equation i.e., $x = \phi(x)$. Then it can be written as

$$\alpha - x_i = j(\alpha - x_{i-1}) \tag{2}$$

$$\alpha - x_{i+1} = j(\alpha - x_i) \tag{3}$$

where j is a contant, $|\phi'(x)| \leq j \leq 1$ for all values of i.

Division of equation (2) by equation (3), resulting in

$$\frac{\alpha - x_i}{\alpha - x_{i+1}} = \frac{\alpha - x_{i-1}}{\alpha - x_i} \tag{4}$$

$$\text{Hence, } \alpha = x_{i+1} - \frac{(x_{i+1} - x_i)^2}{(x_{i+1} - 2x_i - x_{i-1})} \tag{5}$$

But in the successive approximations,

$$\Delta x_i = x_{i+1} - x_i \tag{6}$$

and

$$\Delta^2 x_i = \Delta(\Delta x_i) = \Delta(x_{i+1} - x_i) = \Delta x_{i+1} - \Delta x_i = (x_{i+2} - x_{i+1}) - (x_{i+1} - x_i) = x_{i+2} - 2x_{i+1} + x_i \tag{7.1}$$

$$\Delta^2 x_{i-1} = x_{i+1} - 2x_i + x_{i-1} \tag{7.2}$$

From equation (6), (7.1) and (7.2), yields successive approximation to the root α.

$$\alpha = x_{i+1} - \frac{(\Delta x_i)^2}{\Delta^2 x_{i-1}}$$

Successive Approximations	Δ	Ä 2
x_{i-1}		
	Δx_{i-1}	
x_i		$\Delta^2 x_{i-1}$
	Δx_i	
x_{i+1}		

Example 3.16: Find the root of a function cos x = 3x – 1 through Aitken's Δ^2 method.

Solution: f(x) = cos x – 3x + 1 = 0

For the values of x = 0, then function f(0) = 1, which is +ve

For the value of x = $\frac{\pi}{2}$, then function $f(\frac{\pi}{2}) = -3\ \frac{\pi}{2} + 1 = -8.42857$, which is –ve

Hence the root lies in between 0 and $\frac{\pi}{2}$.

The given equation can be written as; $x = \frac{1}{3}\ (\cos x + 1) = \phi(x)$

Hence, $\phi'(x) = \frac{\sin x}{3}$; and $|\phi'(x)| = \frac{1}{3}|\sin x| < 1$

This represents that iteration can be applied. Consider the initial $x_0 = 0$, then the successive approximations are,

$$x_1 = \phi(x_0) = \frac{1}{3}(\cos 0 + 1) = 0.6667$$

$$x_2 = \phi(x_1) = \frac{1}{3}(\cos 0.667 + 1) = 0.5953$$

$$x_3 = \phi(x_2) = \frac{1}{3}(\cos 0.595 + 1) = 0.6093$$

Successive Approximations	Δ	Δ^2
$x_{i-1} = x_1 = 0.6667$		
	$\Delta x_{i-1} = \Delta x_1 = x_2 - x_1 = -0.0714$	
$x_i = x_2 = 0.5953$		$\Delta^2 x_{i-1} = \Delta x_i - \Delta x_{i-1} = 0.0854$
	$\Delta x_i = \Delta x_2 = x_3 - x_2 = 0.014$	
$x_{i+1} = x_3 = 0.6093$		

Hence the root of a function, cos x = 3x – 1, is 0.607.

3.13 Comparison of Iterative Methods

The Bisection method and Regula-Falsi always converge to an answer, provided a root exist in (a, b) interval and on every iteration the interval width is reduced until the solution is obtained. The Bisection method converges always, but may fail when the function is tangent to the axis and does not cross the x-axis at f(x)=0.. The Newton-Raphson method and the Successive Approximation method require 1 guess and further iterations represents the solution is exact or not. The Bisection method, Regula-Falsi method and the Successive Approximation method (converges when $|\varphi'(x)|<1$ satisfied) converge linearly but Newton-Raphson method converges quadratically with having drawback when derivative is zero.

S.No.	Method for finding solution of non-linear equations	Formula	Order of convergence	Evaluation of function at each step
1	Bisection	$x_{s1} = \frac{a+b}{2}$	One bit	1
2	Regula-Falsi	$x_s = \frac{a\,f(b) + b\,f(a)}{f(b) - f(a)}$	1	1
3	Newton-Raphson	$x_{i+1} = x_i - \frac{f(x_i)}{f'(x_i)}$	2	2
4	Modified Newton-Raphson	$x_{i+1} = x_n - \frac{f_n}{f'\left(x_n - \frac{1}{2}\frac{f_n}{f'_n}\right)}$	3	3
5	Successive Approximation	$x_1 = \varphi(x_0)$	1	1
6	Secant	$x_{i+1} = x_i - \frac{f(x_i)(x_i - x_{i-1})}{f(x_i) - f(x_{i-1})}$	1.62	1

Contd.

7	Muller	$x_{n+1} = x_n + (x_n - x_{n-1})\lambda$	1.84	1
8	Chebyshev	$x_{n+1} = x_n - \frac{f_n}{f'_n} - \frac{1}{2}\frac{f_n^2}{f'^2_n} f''_n$	3	3

3.14 Solution of Non-Linear Equations

The systems of non-linear equations arises in large number of ways, hence method is required to solve them. For simplicity, consider the case of two equations in two unknowns as

f(x, y) = 0 and g(x, y) = 0

In this section, two methods i.e., the iteration method and Newton-Raphson method are discussed.

3.14.1 Iteration Method

Consider the equation be given by,

$$f(x, y) = 0 \tag{1}$$

$$g(x, y) = 0 \tag{2}$$

whose roots are required within a specified accuracy. As in the Iteration Method in section 3.6 for a single equation, hence accordingly the equations (1) and (2) may be written in the form as

$$x = F(x, y) \tag{3}$$

$$y = G(x, y) \tag{4}$$

where the functions F and G satisfy the conditions in the neighbourhood of root as

$$\left|\frac{\partial F}{\partial x}\right| + \left|\frac{\partial F}{\partial y}\right| < 1 \tag{5}$$

$$\left|\frac{\partial G}{\partial x}\right| + \left|\frac{\partial G}{\partial y}\right| < 1 \tag{6}$$

Suppose (x_0, y_0) are the initial approximation to the root (l, m) of the system of equations (1) and (2). After that there is need to construct the successive approximations according to the following formulae as equation (7),

$$x_1 = F(x_0, y_0);\ y_1 = G(x_0, y_0)$$

$$x_2 = F(x_1, y_1);\ y_2 = G(x_1, y_1)$$

....

$x_{n+1} = F(x_n, y_n);\ y_{n+1} = G(x_n, y_n)$

Recently computed values of x_i may be used to compute y_i for faster convergence. On convergence, the result in the limit is

$$j = F(j, k) \tag{8}$$

$$k = G(j, k) \tag{9}$$

Hence l and m are the roots of the system of non-linear equations and the sufficient condition for convergence is equations (5) and (6).

Suppose x = l and y = m is one pair of roots of system of non-linear equations in the closed neighborhood R, if the functions F and G & their 1st partial derivatives are continuous in R; equations (5) and (6) for all (x, y) in R; and the initial approximation (x_0, y_0) is chosen in R then the sequence of approximation is given by equation (7) and converges to the roots x = l and y = m of the system of non-linear equations.

Example 3.17: Find a real root of the following equations

$x = 0.2\,x^2 + 0.8$, and $y = 0.3\,xy^2 + 0.7$

Solution: $F(x, y) = 0.2\,x^2 + 0.8$

$G(x, y) = 0.3\,xy^2 + 0.7$

$$\frac{\partial F}{\partial x} = 0.4x\ ;\ \frac{\partial F}{\partial y} = 0\ ;\ \frac{\partial G}{\partial x} = 0.3y^2\ ;\ \frac{\partial G}{\partial y} = 0.6xy$$

At x = 1 and y = 1, the convergence condition equations (5) and (6) are satisfied, hence are the roots of this system.

Suppose the initial approximation is $x_0 = y_0 = 0.5$, and the convergence condition

$\left|\frac{\partial F}{\partial x}\right| + \left|\frac{\partial F}{\partial y}\right| = 0.2 < 1$ and $\left|\frac{\partial G}{\partial x}\right| + \left|\frac{\partial G}{\partial y}\right| = 0.225 < 1$, is satisfied. Hence the 1st approximation is

$x_1 = F(x_0, y_0) = 0.2\,x_0^2 + 0.8 = 0.85$

$y_1 = G(x_0, y_0) = 0.3\,x_0\,y_0^2 + 0.7 = 0.74$

2nd approximation is

$x_2 = F(x_1, y_1) = 0.2\,x_1^2 + 0.8 = 0.9445$

$y_2 = G(x_1, y_1) = 0.3\,x_1\,y_1^2 + 0.7 = 0.81$

and so on.

It is clear that the convergence of the root (1, 1) is clear. For faster converging rate, put the latest value of x in y, such as, $y_1 = G(x_1, y_0) = 0.2\, x_1^2 + 0.8 = 0.764$, which better approximation than $y_1 = G(x_0, y_0)$.

3.14.2 Newton-Raphson Method

Consider the equation be given by,

$$f(x, y) = 0 \tag{1}$$

$$g(x, y) = 0 \tag{2}$$

whose roots are required within a specified accuracy. Suppose (x_0, y_0) are the initial approximation to the root (l, m) of the system of equations [(1) and (2)] and (h, k) are the difference between exact and approximate value of the root as

$$h = l - x_0 \quad \text{or} \quad l = x_0 + h$$

$$k = m - y_0 \quad \text{or} \quad m = y_0 + k$$

As (l, m) is the exact root, so

$$f(l, m) = 0 \quad \text{i.e.,} \quad f(x_0 + h, y_0 + k) = 0$$

$$g(l, m) = 0 \quad \text{i.e.,} \quad g(x_0 + h, y_0 + k) = 0$$

On expansion of above equations through Taylor's series, resulting

$$f(x_0, y_0) + h\, f_x(x_0, y_0) + k\, f_y(x_0, y_0) + \ldots\ldots = 0$$

$$g(x_0, y_0) + h\, g_x(x_0, y_0) + k\, g_y(x_0, y_0) + \ldots\ldots = 0$$

where f_x, f_y, g_x, g_y are 1st order partial derivative w.r.t. x and y respectively.

Retain the 1st order and neglect the 2nd and higher order of above equations resulting in

$$h\, f_x(x_0, y_0) + k\, f_y(x_0, y_0) = -\, f(x_0, y_0)$$

$$h\, g_x(x_0, y_0) + k\, g_y(x_0, y_0) = -\, g(x_0, y_0)$$

The values of h and k are obtained upon solving the above equations as

$$h = \frac{\begin{vmatrix} -f(x_0, y_0) & f_x(x_0, y_0) \\ -g(x_0, y_0) & g_x(x_0, y_0) \end{vmatrix}}{D} \quad \text{and} \quad k = \frac{\begin{vmatrix} f_x(x_0, y_0) & -f(x_0, y_0) \\ g_x(x_0, y_0) & -g(x_0, y_0) \end{vmatrix}}{D}$$

where $D=\begin{vmatrix} f_x(x_0,y_0) & f_y(x_0,y_0) \\ g_x(x_0,y_0) & g_y(x_0,y_0) \end{vmatrix}$, and upon expansion resulting

$D = f_x(x_0, y_0)\, g_y(x_0, y_0) - g_x(x_0, y_0)\, f_y(x_0, y_0)$

So the new approximation is given through; $x_1 = x_0 + h$ and $y_1 = y_0 + k$

Similarly the new approximation is given through; $x_2 = x_1 + h$ and $y_2 = y_1 + k$; and so on. This process is continued till root of the non-linear equations with desired accuracy are achieved. In the same process, system of 3 or more non-linear equations can be solved.

Consider (x_0, y_0) is the approximation to a root (l, m) of the system of non-linear equations in the closed neighborhood R containing (l, m). If f, g and all their 1st and 2nd order derivatives are continuous and are bounded in R; $x_{i+1} = x_i + h$, and $y_{i+1} = y_i + k$, converges to the root (l, m) of the system.

Example 3.18: Find the root of non-linear system of equations

$x^2 + xy = 6$

$x^2 - y^2 = 3$

Solution: Let $f(x, y) = x^2 + xy - 6 = 0$; and $g(x, y) = x^2 - y^2 - 3 = 0$

x = 2 and y = 1, satisfy the above equations, hence the root are (2, 1).

$$f_x = \frac{\partial f}{\partial x} = 2x + y;\ f_y = \frac{\partial f}{\partial y} = x;\ g_x = \frac{\partial g}{\partial x} = 2x;\ g_y = \frac{\partial g}{\partial y} = -2y$$

Suppose the initial approximation to a root is, $x_0 = 1.9$, $y_0 = 0.9$

Iteration 1: $x_0 = 1.9$, $y_0 = 0.9$

$f(x_0, y_0) = (1.9)^2 + (1.9)(0.9) - 6 = -0.68$

$g(x_0, y_0) = (1.9)^2 - (0.9)^2 - 3 = -0.2$

$f_x = = 2x_0 + y_0 = 2(1.9) + 0.9 = 4.7$

Similarly, $f_y = \frac{\partial f}{\partial y} = 1.9;\ g_x = \frac{\partial g}{\partial x} = 3.8;\ g_y = \frac{\partial g}{\partial y} = -1.8$

$D = f_x(x_0, y_0)\, g_y(x_0, y_0) - g_x(x_0, y_0)\, f_y(x_0, y_0) = (4.7)(-1.8) - (1.9)(3.8) = -15.68$

$$h=\frac{\begin{vmatrix} -f(x_0,y_0) & f_x(x_0,y_0) \\ -g(x_0,y_0) & g_x(x_0,y_0) \end{vmatrix}}{D}=\frac{\begin{vmatrix} 0.68 & 1.9 \\ 0.2 & -1.8 \end{vmatrix}}{-15.68}=\frac{(0.68)(-1.8)-(0.2)(1.9)}{-15.68}=0.102$$

$$k = \frac{\begin{vmatrix} f_x(x_0, y_0) & -f(x_0, y_0) \\ g_x(x_0, y_0) & -g(x_0, y_0) \end{vmatrix}}{D} = \frac{\begin{vmatrix} 4.7 & 0.68 \\ 3.8 & 0.2 \end{vmatrix}}{-15.68} = \frac{(0.47)(1.2) - (3.8)(0.68)}{-15.68} = 0.105$$

Hence, $x_1 = x_0 + h = 1.9 + 0.102 = 2.002$; and $y_1 = y_0 + k = 0.9 + 0.105 = 1.005$

Iteration 2: $x_1 = 2.002$, $y_1 = 1.005$

$f(x_1, y_1) = 0.020$; $g(x_1, y_1) = -0.002$

$$f_x \text{ at } (x_1, y_1) = \frac{\partial f}{\partial x} = 2x_1 + y_1 = 5.009$$

Similarly, $f_y = \dfrac{\partial f}{\partial y} = 2.002$; $g_x = \dfrac{\partial g}{\partial x} = 4.004$; $g_y = \dfrac{\partial g}{\partial y} = -2.01$

$D = f_x(x_1, y_1)\, g_y(x_1, y_1) - g_x(x_1, y_1)\, f_y(x_1, y_1) = (5.009)(-2.01) - (4.004)(2.002)$
$= -18.084$

$$h = \frac{\begin{vmatrix} -f(x_1, y_1) & f_x(x_1, y_1) \\ -g(x_1, y_1) & g_x(x_1, y_1) \end{vmatrix}}{D} = -0.002$$

$$k = \frac{\begin{vmatrix} f_x(x_1, y_1) & -f(x_1, y_1) \\ g_x(x_1, y_1) & -g(x_1, y_1) \end{vmatrix}}{D} = -0.005$$

Hence, $x_2 = x_1 + h = 2.002 + (-0.002) = 2.000$; and $y_2 = y_1 + k = 1.005 + (-0.005) = 1.000$

So, from 1st and 2nd iteration it is clear that there is no change in successive approximation to the root upon 2 decimal places, hence the root of the system of non-linear equations is $x = 2.00$ and $y = 1.00$ upto 2 decimal places.

Exercise

3.1. What are non-linear equations?

3.2. Explain the various methods to solve algebraic and transcendental equations.

3.3. With the help of bisection method:

(a) Find a solution accurate to 4 decimal places for $x = \tan x$ in the interval (4.4, 4.6).

(b) Determine the solution of equation; $8 - 4.5(x - \sin x) = 0$, accurate upto 5 decimal places in the interval (2, 3).

3.4. Use false position method:

(a) Find a solution accurate to within 10-4 for the function $f(x) = x - \cos x$, in the interval $(0, \pi/2)$.

(b) Find a root correct to 3 decimal places for the function, $\tan x - 4x = 0$.

3.5. With the help of fixed point iteration method, find the roots of the equation's correct upto two decimal places;

(a) $f(x) = x^4 + x^2 + 10$

(b) $f(x) = x^2 + 2x + 5$

3.6. A root of $f(x) = x^3 - x^2 - 5 = 0$, lies in the interval (2, 3). Determine this root with Newton-Raphson method for four decimal places.

3.7. A positive root of the equation, $e^x = 1 + x + \frac{x^2}{2} + \frac{x^3}{6} e^{0.3x}$, lies in the interval (2, 3). Use Newton-Raphson method to find this root accurate to five decimal places.

3.8. Use successive approximation method, to find real roots upto 4 significant figures of

(a) $e^{-x} - 10x = 0$ (b) $2x - \log_{10} x - 7 = 0$ (c) $x - \sin x - 0.25 = 0$

3.9. Find a root of the equation's; $x^3 - 3x^2 + 4 = 0$, and $e^x - 2x^2 = 0$; using the modified Newton-Raphson method starting with $x_0 = 1.4$.

3.10. Determine the roots of the following equations through secant method

(a) $x^3 - 75 = 0$, with initial approximation $x_0 = 4$ and $x_1 = 5$

(b) $\cos x . \cosh x - 1 = 0$, with initial approximation $x_0 = 4.5$ and $x_1 = 5.0$

(c) $\sin x - 0.1 x = 0$, with initial approximation $x_0 = 2$ and $x_1 = 3$

3.11. Give the order of convergence of different methods to solve non-linear equations.

3.12. Find a root of the equation's;

(a) $x^3 - 3x^2 + 4 = 0$, given that a root is near 1.0, and

(b) $e^x - 2x^2 = 0$, given that a root is near 4.0,

using the Muller's method.

3.13. Determine the root of the equation's through Aitken's method

(a) $\cos x - x e^x = 0$

(b) $x^3 - 5x - 11 = 0$, correct upto 3 decimal places

(c) $e^x - 3x = 0$. Lying between 0 and 1

(d) $5x^3 - 20x + 3 = 0$

3.14. Compare various iteration methods.

4

Numerical Differentiation

4.1 Introduction

Numerical differentiation is the process of computing the function's derivative values from the given set of numerical values when the actual form of the function is not known.

Similar to the numerical interpolation, a number of formulae for differentiation are derived in this chapter and are

(1) Derivatives Based on Newton's Forward Interpolation Formula: This is used to compute the derivative for some given x lying near the beginning of the data table.

(2) Derivatives Based on Newton's Backward Interpolation Formula: This is used to compute the derivative for a point near the end of the data table.

(3) Derivatives Based on Stirling's and Bessel's Interpolation Formula: This is used to compute the derivative for some point lying in the middle of the data table.

(4) Derivatives Based on Newton's Divided Difference Formula: This is used to compute the derivative for some point when the data table is not equi-spaced.

4.2 Derivatives Based on Newton's Forward Interpolation Formula

Suppose the function $y = f(x)$ is known at $(n+1)$ equispaced points $x_0, x_1, \ldots\ldots\ldots,$ x_n and they are $y_0, y_1, \ldots\ldots\ldots, y_n$ respectively i.e., $y_i = f(x_i)$, $i=0, 1, \ldots\ldots\ldots, n$.

$x = x_0 + ph$; p is any real number and h is equispacing,

Newton's forward difference interpolation formula is

$$y = y_0 + \frac{p}{1!}\Delta y_0 + \frac{p\,(p-1)}{2!}\Delta^2 y_0 + \frac{p\,(p-1)\,(p-2)}{3!}\Delta^3 y_0 + \frac{p\,(p-1)\,(p-2)\,(p-3)}{4!}\Delta^4 y_0 + \ldots\ldots \tag{1}$$

On considering upto four terms and differentiate it with p yielding

$$\frac{dy}{dp} = \Delta y_0 + \frac{2p-1}{2!}\Delta^2 y_0 + \frac{3p^2-6p+2}{3!}\Delta^3 y_0 + \frac{4p^3-18p^2+22p-6}{3!}\Delta^4 y_0 + \cdots \quad (2)$$

Since, $p = \frac{x - x_0}{h}$, therefore $\frac{dp}{dx} = \frac{1}{h}$

Now

$$\frac{dy}{dx} = \frac{dy}{dp} X \frac{dp}{dx} \text{ (By Chain Rule)}$$

$$\frac{dy}{dx} = \frac{dy}{dp} X \frac{dp}{dx} = \frac{1}{h}\left[\Delta y_0 + \frac{2p-1}{2!}\Delta^2 y_0 + \frac{3p^2-6p+2}{3!}\Delta^3 y_0 + \frac{4p^3-18p^2+22p-6}{3!}\Delta^4 y_0 + \cdots\cdots\right] \quad (3)$$

At $x = x_0$, $p = 0$ yielding

$$\left(\frac{dy}{dx}\right)_{x_0} = \frac{1}{h}\left[\Delta y_0 - \frac{1}{2}\Delta^2 y_0 + \frac{1}{3}\Delta^3 y_0 - \frac{1}{4}\Delta^4 y_0 + \frac{1}{5}\Delta^5 y_0 - \frac{1}{6}\Delta^6 y_0 \cdots\cdots\right]$$

Differentiate equation (3) w.r.t x, yielding

$$\frac{d^2y}{dx^2} = \frac{d}{dp}\left(\frac{dy}{dp}\right)\frac{dp}{dx} = \frac{1}{h^2}\left[\frac{2}{2!}\Delta^2 y_0 + \frac{6p-6}{3!}\Delta^3 y_0 + \frac{12p^2-36p+22}{3!}\Delta^4 y_0 + \cdots\cdots\right] \quad (4)$$

At $x = x_0$, $p = 0$, the above equation yields

$$\left(\frac{d^2y}{dx^2}\right)_{x_0} = \frac{1}{h^2}\left[\Delta^2 y_0 - \Delta^3 y_0 + \frac{11}{12}\Delta^4 y_0 - \frac{5}{6}\Delta^5 y_0 + \frac{137}{180}\Delta^6 y_0 \cdots\cdots\right]$$

Differentiate equation (4) again w.r.t x, yielding

$$\frac{d^3y}{dx^3} = \frac{d}{dp}\left(\frac{d^2y}{dx^2}\right)\frac{dp}{dx} = \frac{1}{h^3}\left[\Delta^3 y_0 + \frac{2p-3}{2}\Delta^4 y_0 + \cdots\cdots\right] \quad (5)$$

At $x = x_0$, $p = 0$, the above equation yields

$$\left(\frac{d^3y}{dx^3}\right)_{x_0} = \frac{1}{h^3}\left[\Delta^3 y_0 - \frac{3}{2}\Delta^4 y_0 + \cdots\cdots\right]$$

Example 4.1: Find the first and second order derivatives of the function y = f(x) tabulated below at the point x = 3.

x	3	3.2	3.4	3.6	3.8	4.0
y = f(x)	-14.000	-10.032	-5.296	-0.256	6.672	14.000

Solution: As derivatives are required at x=3 which is at the beginning of the table, so use Newton's forward formula.

The difference table is

x	Y	Δy	$\Delta^2 y$	$\Delta^3 y$	$\Delta^4 y$	$\Delta^5 y$
3	-14.000 (y_0)					
		3.968 (Δy_0)				
3.2	-10.032		0.768 ($\Delta^2 y_0$)			
		4.736		0.048 ($\Delta^3 y_0$)		
3.4	-5.296		0.816		0	
		5.552		0.048		0
3.6	-0.256		0.864		0	
		6.416		0.048		
3.8	6.672		0.912			
		7.328				
4.0	14.000					

Here $x_0 = 3$, h = 0.2, x = 3

By Newton's forward formula

$$\left(\frac{dy}{dx}\right)_{x_0} = \frac{1}{h}\left[\Delta y_0 - \frac{1}{2}\Delta^2 y_0 + \frac{1}{3}\Delta^3 y_0 - \frac{1}{4}\Delta^4 y_0 + \frac{1}{5}\Delta^5 y_0 - \frac{1}{6}\Delta^6 y_0 \ldots\ldots\ldots\ldots\right]$$

and

$$\left(\frac{d^2y}{dx^2}\right)_{x_0} = \frac{1}{h^2}\left[\Delta^2 y_0 - \Delta^3 y_0 + \frac{11}{12}\Delta^4 y_0 - \frac{5}{6}\Delta^5 y_0 + \frac{137}{180}\Delta^6 y_0 \ldots\ldots\ldots\ldots\right]$$

Therefore

$$\left(\frac{dy}{dx}\right)_{x=3} = \frac{1}{0.2}\left[3.968 - \frac{1}{2}(0.768) + \frac{1}{3}(0.048)\right] = 18$$

and $$\left(\frac{d^2y}{dx^2}\right)_{x=3} = \frac{1}{(0.2)^2}\left[0.768 - 0.048\right] = 18$$

4.3 Derivatives Based on Newton's Backward Interpolation Formula

Suppose the function y = f(x).

where $x = x_n + ph$; p is any real number and h is equispacing,

$$p = \frac{x - x_n}{h}, \text{ therefore } \frac{dp}{dx} = \frac{1}{h}$$

Newton's backward difference interpolation formula is

$$y = yn + \frac{p}{1!}\nabla y_n + \frac{p(p+1)}{2!}\nabla^2 y_n + \frac{p(p+1)(p+2)}{3!}\nabla^3 y_n +$$
$$\frac{p(p+1)(p+2)(p+3)}{4!}\nabla^4 y_n + \frac{p(p+1)\,.........\,(p+n-1)}{n!}\nabla^n y_n \quad \text{-----(1)}$$

Differentiate it with p yielding

$$\frac{dy}{dp} = \nabla y_n + \frac{2p+1}{2!}\nabla^2 y_n + \frac{3p^2+6p+2}{3!}\nabla^3 y_n + \frac{4p^3+18p^2+22p+6}{4!}\nabla^4 y_n + ..(2)$$

Now

$$\frac{dy}{dx} = \frac{dy}{dp} X \frac{dp}{dx} \text{ (By Chain Rule)}$$

$$\frac{dy}{dx} = \frac{dy}{dp} X \frac{dp}{dx} = \frac{1}{h}\left[\nabla y_n + \frac{2p+1}{2!}\nabla^2 y_n + \frac{3p^2+6p+2}{3!}\nabla^3 y_n + \frac{4p^3+18p^2+22p+6}{4!}\nabla^4 y_n + ...\right]$$
$$\text{...... (3)}$$

At $x = x_n$, $p = 0$ yielding

$$\left(\frac{dy}{dx}\right)_{x_n} = \frac{1}{h}\left[\nabla y_n + \frac{1}{2}\nabla^2 y_n + \frac{1}{3}\nabla^3 y_n + \frac{1}{4}\nabla^4 y_n + \frac{1}{5}\nabla^5 y_n + \frac{1}{6}\nabla^6 y_n + ...\right]$$

Differentiate equation (3) w.r.t x, yielding

$$\frac{d^2y}{dx^2} = \frac{d}{dp}\left(\frac{dy}{dx}\right) X \frac{dp}{dx} = \frac{1}{h^2}\left[\nabla^2 y_n + \frac{6p+6}{3!}\nabla^3 y_n + \frac{6p^2+18p+11}{4!}\nabla^4 y_n + ...\right] \text{--(4)}$$

At $x = x_n$, $p = 0$ yielding

$$\left(\frac{d^2y}{dx^2}\right)_{x_n} = \frac{1}{h^2}\left[\nabla^2 y_n + \Delta^3 y_n + \frac{11}{12}\Delta^4 y_n + \frac{5}{6}\Delta^5 y_n + \frac{137}{180}\Delta^6 y_n \,.........\right]$$

Differentiate equation (4) again w.r.t x, yielding

$$\frac{d^3y}{dx^3} = \frac{d}{dp}\left(\frac{d^2y}{dx^2}\right)\frac{dp}{dx} = \frac{1}{h^3}\left[\nabla^3 y_0 + \frac{2p-3}{2}\nabla^4 y + ...\right] \text{------------------------(5)}$$

At $x = x_0$, $p = 0$, the above equation yields

$$\left(\frac{d^3y}{dx^3}\right)_{x_n} = \frac{1}{h^3}\left[\nabla^3 y_0 - \frac{3}{2}\nabla^4 y_0 + ...\right]$$

Example 4.2: Find first derivative at x = 0.4 from the following table

x	0.1	0.2	0.3	0.4
y	1.10517	1.22140	1.34986	1.49182

Solution: Since x = 0.4 which is at the end of table, so use Newton's backward formula.

Here we have h=0.1, x_n=0.4 and x=0.4

$$p = \frac{x - x_n}{h} = 0$$

The difference table is

X	y	10^5 y	$10^5\,\Delta y$	$10^5\,\Delta^2 y$	$10^5\,\Delta^3 y$
0.1	1.10517	110517			
			11623		
0.2	1.22140	122140		1223	
			12846		127
0.3	1.34986	134986		1350	
			14196		
0.4	1.49182	149182			

By Newton's backward formula

$$\left(\frac{dy}{dx}\right)_{x_n} = \frac{1}{h}\left[\nabla y_n + \frac{1}{2}\nabla^2 y_n + \frac{1}{3}\nabla^3 y_n + \dots\dots\dots\dots\right]$$

At x = 0.4,

$$10^5 * \left(\frac{dy}{dx}\right)_{x=0.4} = \frac{1}{0.1}\left[14196 + \frac{1350}{2} + \frac{127}{3}\right] = 149133$$

Example 4.3: A rod is rotating in a plane about one of its end. If the following table gives the angle θ radians through which the rod has turned for different values of time t seconds. Find its angular velocity ad angular acceleration at t = 0.7 seconds.

t (in seconds)	0	0.2	0.4	0.6	0.8	1.0
θ (in radians)	0	0.12	0.48	0.10	2.0	3.20

Solution: Since t = 0.7 which is near the end of table, so use Newton's backward formula.

Here we have h=0.2, t_n=0.4 and t=0.7

$$p = \frac{t - t_n}{h} = -1.5$$

The difference table is

T	θ	Δθ	$\Delta^2\theta$	$\Delta^3\theta$	$\Delta^4\theta$	$\Delta^5\theta$
0	0					
		0.12				
0.2	0.12		0.24			
		0.36		0.02		
0.4	0.48		0.26		0	
		0.62		0.02		0
0.6	0.10		0.28		0	
		0.90		0.02		
0.8	2.0		0.30			
		1.20				
1.0	3.20					

By Newton's backward formula

$$\frac{d\theta}{dt} = \frac{1}{h}\left[\nabla\theta_n + \frac{2p+1}{2!}\nabla^2\theta_n + \frac{3p^2+6p+2}{3!}\nabla^3\theta_n + \frac{4p^3+18p^2+22p+6}{4!}\nabla^4\theta_n + ...\right]$$

and $$\frac{d^2\theta}{dt^2} = \frac{1}{h^2}\left[\nabla^2\theta_n + \frac{6p+6}{3!}\nabla^3\theta_n + \frac{6p^2+18p+11}{4!}\nabla^4\theta_n + ...\right]$$

At $t = 0.7$,

$$\left(\frac{d\theta}{dx}\right)_{t=0.7} = \frac{1}{0.2}\left[1.20 + \frac{2(-1.5)+1}{2!}(0.30) + \frac{3(-1.5)^2+6(-1.5)+2}{3!}(0.02)\right]$$

= 4.496 radians/seconds

So, angular velocity is 4.496 radians per second.

$$\left(\frac{d^2\theta}{dt^2}\right)_{t=0.7} = \frac{1}{(0.2)^2}[\,0.30-(0.5)*(0.02)] = 7.25 \text{ radians per second}$$

So, angular acceleration is 7.25 radians square per second.

4.4 Derivatives Based on Stirling's Interpolation Formula

Suppose the function $y = f(x)$.

where $x = x_0 + ph$; p is any real number and h is equispacing,

$$p = \frac{x - x_0}{h}, \text{ therefore } \frac{dp}{dx} = \frac{1}{h}$$

Stirling's formula is

$$y_p = y_0 + p\left[\frac{\Delta y_0 + \Delta y_{-1}}{2}\right] + \frac{p^2}{2!}\Delta^2 y_{-1} + \frac{p(p^2-1)}{3!}\left[\frac{\Delta^3 y_{-1} + \Delta^3 y_{-2}}{2}\right]$$

$$+\frac{p^2(p^2-1)}{4!}\Delta^4 y_{-2} + \frac{p(p^2-1)(p^2-2^2)}{5!}\left[\frac{\Delta^5 y_{-2} + \Delta^5 y_{-3}}{2}\right] + \quad \text{-------(1)}$$

Differentiate equation (1) w.r.t x, yielding

$$\frac{dy}{dx} = \frac{dy}{dp}\frac{dp}{dx}$$

$$\frac{dy}{dx} = \frac{1}{h}\left[\begin{array}{l}\left[\frac{\Delta y_0 + \Delta y_{-1}}{2}\right] + p\,\Delta^2 y_{-1} - \frac{(3p^2-1)}{6}\left[\frac{\Delta^3 y_{-1} + \Delta^3 y_{-2}}{2}\right] + \frac{(4p^3-2p)}{24}\Delta^4 y_{-2} + \\ \frac{(5p^4-15p^2+4)}{120}\left[\frac{\Delta^5 y_{-2} + \Delta^5 y_{-3}}{2}\right] + \ldots\ldots\ldots\ldots\end{array}\right]$$

-------------------(2)

At $x=x_0$ and $p=0$, yielding

$$\left(\frac{dy}{dx}\right)_{x_0} = \frac{1}{h}\left[\left(\frac{\Delta y_0 + \Delta y_{-1}}{2}\right) - \frac{1}{6}\left(\frac{\Delta^3 y_{-1} + \Delta y^4_{-2}}{2}\right) + \frac{1}{30}\left(\frac{\Delta^5 y_{-2} + \Delta y^5_{-3}}{2}\right) + \ldots\ldots\right]$$

Differentiate equation (2) w.r.t x, yielding

$$\frac{d^2y}{dx^2} = \frac{d}{dp}\left(\frac{dy}{dx}\right)\frac{dp}{dx}$$

$$\frac{d^2y}{dx^2} = \frac{1}{h^2}\left[\Delta^2 y_{-1} + p\left(\frac{\Delta^3 y_{-1} + \Delta^3 y_{-2}}{2}\right) + \frac{(6p^2-1)}{12}\Delta y^4_{-2} + \left(\frac{\Delta^3 y_{-1} + \Delta^4 y_{-2}}{2}\right) + \frac{(2p^3-3p)}{12}\left(\frac{\Delta^5 y_{-2} + \Delta^5 y_{-3}}{2}\right) + \ldots\right]$$

------------------(3)

At $x=x_0$ and $p=0$, yielding

$$\left(\frac{d^2y}{dx^2}\right)_{x_0} = \frac{1}{h^2}\left[\Delta^2 y_{-1} - \frac{1}{12}\Delta^4 y_{-2} + \frac{1}{90}\Delta^6 y_{-3} + \ldots\ldots\right]$$

Example 4.4: A rotor in a turbine moves along a fixed rod. Its distance x cm along the rod is given for various values of the time t seconds. Find the velocity of the slider and its acceleration when t = 0.3 seconds.

T	0.0	0.1	0.2	0.3	0.1	0.5	0.6
Y	30.13	31.62	32.87	33.64	33.95	33.81	33.24

Solution: As the derivatives are required at t=0.3 which exist in the middle of the table, so use Stirling's formula for differentiation.

Take h = 0.1, $t_0 = 0.3$

Here we have h=0.2, t_n=0.4 and t=0.7

$$p = \frac{t - t_n}{h} = -1.5$$

The difference table is

t	p	Y	Δy	$\Delta^2 y$	$\Delta^3 y$	$\Delta^4 y$	$\Delta^5 y$	$\Delta^6 y$
0.0	-3	30.13						
			1.49					
0.1	-2	31.62		-0.24				
			1.25		-0.24			
0.2	-1	32.87		-0.48		0.26		
			0.77 (Δy_{-1})		0.02 $(\Delta^3 y_{-2})$		-0.27 $(\Delta^5 y_{-3})$	
0.3	0	33.64		-0.46 $(\Delta^2 y_{-1})$		-0.01 $(\Delta^4 y_{-2})$		0.29 $(\Delta^6 y_{-3})$
			0.31 (Δy_0)		0.01 $(\Delta^3 y_{-1})$		0.02 $(\Delta^5 y_{-2})$	
0.4	1	33.95		-0.45		0.1		
			-0.14		0.02			
0.5	2	33.81		-0.43				
			-0.57					
0.6	3	33.24						

By Stirling's formula

$$\frac{dy}{dx} = \frac{1}{h}\left[\left[\frac{\Delta y_0 + \Delta y_{-1}}{2}\right] + p\,\Delta^2 y_{-1} - \frac{(3p^2-1)}{6}\left[\frac{\Delta^3 y_{-1} + \Delta^3 y_{-2}}{2}\right] + \frac{(4p^3-2p)}{24}\Delta^4 y_{-2} + \frac{(5p^4 - 15p^2 + 4)}{120}\left[\frac{\Delta^5 y_{-2} + \Delta^5 y_{-3}}{2}\right] + \ldots\ldots\ldots\ldots\right]$$

and

$$\frac{d^2y}{dx^2}=\frac{1}{h^2}\left[\Delta^2 y_{-1}+p\left(\frac{\Delta^3 y_{-1}+\Delta^3 y_{-2}}{2}\right)+\frac{(6p^2-1)}{12}\Delta y^4_{-2}+\left(\frac{\Delta^3 y_{-1}+\Delta^4 y_{-2}}{2}\right)+\frac{(2p^3-3p)}{12}\left(\frac{\Delta^5 y_{-2}+\Delta^5 y_{-3}}{2}\right)+......\right]$$

At $t=t_0=0.3$

$$\left(\frac{dy}{dx}\right)_{t_0}=\frac{1}{h}\left[\left(\frac{\Delta y_0+\Delta y_{-1}}{2}\right)-\frac{1}{6}\left(\frac{\Delta^3 y_{-1}+\Delta y^4_{-2}}{2}\right)+\frac{1}{30}\left(\frac{\Delta^5 y_{-2}+\Delta y^5_{-3}}{2}\right)\right]$$

Therefore

$$\left(\frac{dy}{dx}\right)_{t=0.3}=\frac{1}{0.1}\left[\left(\frac{0.31+0.77}{2}\right)-\frac{1}{6}\left(\frac{0.01+0.02}{2}\right)+\frac{1}{30}\left(\frac{0.02-0.27}{2}\right)\right]=5.33$$

Velocity of the slider at t=0.3 seconds is 5.33 cm/sec.

and $$\left(\frac{d^2y}{dx^2}\right)_{t_0}=\frac{1}{h^2}\left[\Delta^2 y_{-1}-\frac{1}{12}\Delta^4 y_{-2}+\frac{1}{90}\Delta^6 y_{-3}\right]$$

Therefore $$\left(\frac{d^2y}{dx^2}\right)_{t=0.3}=\frac{1}{(0.1)^2}\left[(-0.46)-\frac{1}{12}(-0.01)+\frac{1}{90}(0.29)\right]=-45.60$$

Acceleration of the slider at t=0.3 seconds is -45.66 cm/sec^2.

4.5 Derivatives Based on Bessel's Interpolation Formula

Suppose the function $y = f(x)$.

where $x = x_0 + ph$; p is any real number and h is equispacing,

$$p=\frac{x-x_0}{h},\ \text{therefore}\ \frac{dp}{dx}=\frac{1}{h}$$

Bessel's formula is

$$y=y_0+\frac{1}{2}(y_0+y_1)+\left(p-\frac{1}{2}\right)\Delta y_0+\frac{p(p-1)}{2!}.\frac{1}{2}\left(\Delta^2 y_{-1}+\Delta^2 y_0\right)+\frac{\left(p-\frac{1}{2}\right)p(p-1)}{3!}\Delta^3 y_{-1}$$
$$+\frac{(p+1)p(p-1)(p-2)}{4!}.\frac{1}{2}\left(\Delta^4 y_{-1}+\Delta^4 y_0\right)+\ ...$$

$$y=y_0+p\,\Delta y_0+\frac{p(p-1)}{2!}.\frac{1}{2}\left(\Delta^2 y_{-1}+\Delta^2 y_0\right)+\frac{\left(p-\frac{1}{2}\right)p(p-1)}{3!}\Delta^3 y_{-1}$$
$$+\frac{(p+1)p(p-1)(p-2)}{4!}.\frac{1}{2}\left(\Delta^4 y_{-1}+\Delta^4 y_0\right)+ \qquad \text{------(1)}$$

Differentiate equation (1) w.r.t p, yielding

$$\frac{dy}{dp} = \Delta y_0 + \frac{2p-1}{2!}\cdot\frac{1}{2}\left(\Delta^2 y_{-1} + \Delta^2 y_0\right) + \frac{\left(3p^2 - 3p + \frac{1}{2}\right)}{3!}\Delta^3 y_{-1}$$

$$+\frac{4p^3 - 6p^2 - 2p + 2}{4!}\cdot\frac{1}{2}\left(\Delta^4 y_{-1} + \Delta^4 y_0\right) + \qquad \text{-------- (2)}$$

Now

$$\frac{dy}{dx} = \frac{dy}{dp}\cdot\frac{dp}{dx} \text{ (By Chain Rule)}$$

$$\frac{dy}{dx} = \frac{1}{h}\left[\Delta y_0 + \frac{2p-1}{2!}\left(\frac{\Delta^2 y_{-1} + \Delta^2 y_0}{2}\right) + \frac{\left(3p^2 - 3p + \frac{1}{2}\right)}{3!}\Delta^3 y_{-1} + \frac{(4p^3 - 6p^2 - 2p + 2)}{4!}\left(\frac{\Delta^4 y_{-2} + \Delta^4 y_{-1}}{2}\right)\right] \qquad \text{........ (3)}$$

At $x = x_0$, $p = 0$ yielding

$$\left(\frac{dy}{dx}\right)_{x_0} = \frac{1}{h}\left[\Delta y_0 - \frac{1}{4}\left(\frac{\Delta^2 y_{-1} + \Delta^2 y_0}{2}\right) + \frac{1}{12}\Delta^3 y_{-1} + \frac{1}{24}\left(\frac{\Delta^4 y_{-2} + \Delta^4 y_{-1}}{2}\right)\right]$$

Differentiate equation (3) w.r.t x, yielding

$$\frac{d^2y}{dx^2} = \frac{d}{dp}\left(\frac{dy}{dp}\right)\frac{dp}{dx} = \frac{1}{h^2}\left[\frac{(\Delta^2 y_{-1} + \Delta^2 y_0)}{2} + \frac{6p-6}{3!}\Delta^3 y_{-1} + \frac{12p^2 - 12p - 2}{4!}\frac{(\Delta^4 y_{-2} + \Delta^4 y_{-1})}{2}\right] \qquad \text{--------- (4)}$$

At $x = x_0$, $p = 0$, the above equation yields

$$\left(\frac{d^2y}{dx^2}\right)_{x_0} = \frac{1}{h^2}\left[\frac{(\Delta^2 y_{-1} + \Delta^2 y_0)}{2} - \frac{1}{2}\Delta^3 y_{-1} - \frac{1}{24}(\Delta^4 y_{-2} + \Delta^4 y_{-1})\right]$$

Example 4.5: Find the value at 1st derivative at x=4 through the below mentioned table

x	1	2	3	4	5	6	7
y = f(x)	0	1	3	5	8	12	18

Solution: As the derivatives are required at x=4 which exist in the middle of the table, so use Bessel's formula for differentiation.

Here h = 1, x = 4

The difference table is

t	p	Y	Δy	$\Delta^2 y$	$\Delta^3 y$	$\Delta^4 y$	$\Delta^5 y$	$\Delta^6 y$
1	-3	0						
			1					
2	-2	1		1				
			2		-1			
3	-1	3		0		2		
			$2(\Delta y_{-1})$		$1(\Delta^3 y_{-2})$		$-3\ (\Delta^5 y_{-3})$	
4	0	5		$1\ (\Delta^2 y_{-1})$		$-1\ (\Delta^4 y_{-2})$		$5(\Delta^6 y_{-3})$
			$3\ (\Delta y_0)$		$0(\Delta^3 y_{-1})$		$2\ (\Delta^5 y_{-2})$	
5	1	8		$1\ (\Delta^2 y_0)$		$1\ (\Delta^4 y_{-1})$		$0(\Delta^6 y_{-2})$
			$4\ (\Delta y_1)$		1			
6	2	12		2				
			6					
7	3	18						

Using Bessel's formula

$$\left(\frac{dy}{dx}\right)_{x_0} = \frac{1}{h}\left[\Delta y_0 - \frac{1}{4}(\Delta^2 y_{-1} + \Delta^2 y_0) + \frac{1}{12}\Delta^3 y_{-1} + \frac{1}{24}(\Delta^4 y_{-2} + \Delta^4 y_{-1}) - \right.$$

$$\left. \frac{1}{120}\Delta^5 y_{-2} - \frac{1}{240}(\Delta^6 y_{-3} + \Delta^6 y_{-2})\right]$$

$$\left(\frac{dy}{dx}\right)_{x=4} = \frac{1}{1}\left[3 - \frac{1}{4}(1+1) + \frac{1}{12}(0) + \frac{1}{24}(-1+1) - \frac{1}{120}(2) - \frac{1}{240}(5+0)\right] = 2.4625$$

4.6 Derivatives Based on Newton's Divided Difference Formula

Newton's divided difference formula is

$f(x)=y = y_0 + (x - x_0)\,[x_0, x_1] + (x - x_0)(x - x_1)\,[x_0, x_1, x_2] + (x - x_0)(x - x_1)(x - x_2)\,[x_0, x_1, x_2, x_3] + \dots + (x - x_0)(x - x_1) \dots (x - x_n)\,[x, x_0, \dots x_n]$ ....(1)

where $[x_0, x_1]$ represents $f(x_0, x_1)$

Differentiate equation (1) w.r.t x, yielding

$f'(x) = f(x_0,x_1) + \{(x - x_0)(x - x_1)\}\, f(x_0, x_1, x_2) + \{(x - x_0)(x - x_1) + (x - x_0)(x - x_2) + (x - x_1)(x - x_2)\, f(x_0, x_1, x_2, x_3) + \dots$ .... (2)

Differentiate equation (2) w.r.t x, yielding

$f''(x) = 2.f(x_0,x_1,x_2)+\{(x - x_0)+ (x - x_1)+ (x - x_0)+ (x - x_2)+ (x - x_1)+ (x - x_2)\}\ f(x_0,x_1,x_2,x_3) + \dots$

$f''(x) = 2.f(x_0,x_1,x_2)+ 2\{(x - x_0)+ (x - x_1)+ (x - x_2)\}\ f(x_0,x_1,x_2,x_3) + \dots$..... (3)

and so on.

Example 4.6: Find the value at 1st derivative at x=5 through the below mentioned table

x	0	2	3	4	7	9
y = f(x)	40	26	58	112	466	922

Solution: The x values are not equi-spaced, so use Newton's divided difference formula for differentiation.

The difference table is

X	y = f(x)	Δy [Δf(x)]	Δ^2y [Δ^2f(x)]	Δ^3y[Δ^3f(x)]
0 (x_0)	4 (y_0)			
		11		
2 (x_1)	26 (y_1)	$\left(\Delta y_0 = \frac{y_1 - y_0}{x_1 - x_0}\right)$	7	
		32		1
3	58		11	
		54		1
4	112		16	
7	466	118		1
			22	
9	922	228		

3rd divided difference is constant, the formula is

$f(x)= f(x_0) + (x - x_0)\, \Delta f(x_0) + (x - x_0)(x - x_1)\, \Delta^2 f(x_0) + (x - x_0)(x - x_1)(x - x_2)\, \Delta^3 f(x_0)$

Differentiate above equation, yielding

$f'(x)= \Delta f(x_0) + \{(x-x_0)+(x-x_1)\}\Delta^2 f(x_0) + \{(x-x_0)(x-x_1) + (x-x_0)(x-x_2) + (x-x_1)(x-x_2)\}\Delta^3 f(x_0)$ (1)

Put $x_0 = 0$, $x_1 = 2$, $x_2 = 3$, $x = 5$, and values of $\Delta f(x_0)$, $\Delta^2 f(x_0)$, $\Delta^3 f(x_0)$ in the above equation, which yields

$f'(x)= 11 + \{(5-0)+(5-2)\}(7) + \{(5-0)(5-2) + (5-0)(5-3) + (5-2)(5-3)\}(1) = 98$

Exercise

4.1 Find the first and second order derivatives of the function y = f(x) tabulated below at the point x = 1.1 and 1.6.

x	1.0	1.1	1.2	1.3	1.4	1.5	1.6
y = f(x)	7.989	8.403	8.781	9.129	9.451	9.750	10.031

4.2 The following data gives the velocity of a particle for 20 seconds at an interval of 5 seconds. Find the initial acceleration using the entire data:

Time, t (sec)	:	0	5	10	15	20
Velocity, v (m/sec)	:	0	3	14	69	228

4.3 Find the value of cos(1.74) through the following table

X	1.7	1.74	1.78	1.82	1.86
sin x	0.9916	0.9857	0.9781	0.9691	0.9584

4.4 The population of a town (as obtained from census data) is shown in the following table

Year	:	1961	1971	1981	1991	2001
Population (in 10^3)	:	19.96	39.65	58.81	77.21	94.61

4.5 Find dy/dx and d^2y/dx^2 at x = 1.05, 1.25 and 1.15 through the following data

x	:	1.00	1.05	1.10	1.15	1.20	1.25	1.30
y	:	1.000	1.025	1.049	1.072	1.095	1.118	1.140

4.6 The following data gives corresponding values of pressure and specific volume of superheated steam

Sp. Volume (v)	:	2	4	6	8	10
Pressure (p)	:	105	42.7	25.3	16.7	13

Find the rate of change of

(i) Pressure with respect to volume when v = 2

(ii) Volume with respect to pressure when p = 105

4.7 A rod is rotating in a plane. The following table gives the angle (radians) through which the rod has turned for various values of time t second.

Time	:	0	0.2	0.4	0.6	0.8	1.0	1.2
Angle	:	0	0.12	0.49	1.12	2.02	3.20	4.67

Calculate the angular velocity and the angular acceleration of the rod, when t = 0.6 second.

4.8 Find first and second derivatives at x = 0.4 from the following table

x	0.1	0.2	0.3	0.4
y	1.10517	1.22140	1.34986	1.49182

4.9 Compute the values of $y'(1.2)$ and $y''(1.2)$ from the following table

(a)

x	1	1.1	1.2	1.3	1.4
y	0.1	0.34	0.42	0.53	0.62

(b)

x	1	1.1	1.2	1.3	1.4
y	0.0254	0.0437	0.0587	0.0670	0.0780

(b)

x	1	1.1	1.2	1.3	1.4
y	0.0012	0.2342	0.5786	0.7693	0.8934

5

Finite Differences and Interpolation

5.1 Introduction

In applied mathematics the solution of problem generally consists of numbers which satisfy some kind of equation which are generally specified by numbers. But in practice it is found that these equations not always possible to express numerically or in exact decimal representative of the solution. Numerical methods are very important tools to provide practical methods for calculating the solutions of applied mathematics to a desired degree of accuracy. The wide use of electronic computers for solving problems in the fields of engineering, scientific, industry etc that further enhanced the scope of numerical methods.

The calculus of finite differences deals with the changes in the values of the function (dependent variable) due to changes in the independent variables. Again through this the relation of the values of the function is studied whenever the independent variable changes by finite jumps whether equal or unequal.

5.2 Finite Difference Operators

Consider the function $y = f(x)$, then the values of y corresponding to different values of x be a, a+h, a+2h, ..., a+nh differing by h, the values of y=f(x) corresponding to the values of x are f(a), f(a+h), f(a+2h),......, f(a+nh).

The values of the independent variable x are called **argument** and corresponding value of dependent variable y is called **entry**.

5.2.1 Forward Differences

Now, f(a+h)-f(a), f(a+2h)-f(a+h), ..., f(a+nh)-f(a+$\overline{n-1}$h) are called the first difference of the function y=f(x) and it is called the first differences of the function y=f(x) and it is denoted as Δf(a), Δf(a+h), ..., Δf(a+nh) etc, i.e.,

$$\left.\begin{aligned} f(a) &= f(a+h) - f(a) \\ \Delta f(a+h) &= f(a+2h) - f(a+h) \\ &\dots\dots\dots\dots\dots\dots \\ \Delta f(a+nh) &= f(a+nh) - f(a+\overline{n-1}h) \end{aligned}\right\} \quad (a)$$

where Δ is called the **forward difference operator** and $\Delta f(a)$, $\Delta f(a+h)$ are called first forward differences.

In general the first forward difference is defined by $\Delta f(x) = f(x+h) - f(x)$.

The differences of the first forward differences given by equation (a) are called the second forward differences and are denoted by $\Delta^2 f(a)$, $\Delta^2 f(a+h)$ etc.

Therefore, we have

$$\begin{aligned} \Delta^2 f(a) &= \Delta f(a+h) - \Delta f(a) \\ &= [f(a+2h) - f(a+h)] - [f(a+h) - f(a)] \\ &= f(a+2h) - 2f(a+h)] + f(a) \end{aligned}$$

Similarly,

$$\begin{aligned} \Delta^2 f(a+h) &= \Delta f(a+2h) - \Delta f(a+h) \\ &= [f(a+3h) - f(a+2h)] - [f(a+2h) - f(a+h)] \\ &= f(a+3h) - 2f(a+2h)] + f(a+h) \end{aligned}$$

In general we have

$$\begin{aligned} \Delta^2 f(x) &= \Delta f(x+h) - \Delta f(x) \\ &= [f(x+2h) - f(x+h)] - [f(x+h) - f(x)] \\ &= f(x+2h) - 2f(x+h)] + f(x) \end{aligned}$$

Again difference of the second forward differences are called third forward differences and so on. In general the nth forward difference is given by $\Delta^3 f(a)$, $\Delta^3 f(a+h)$ etc, i.e.,

$$\begin{aligned} \Delta^3 f(a) &= \Delta^2 f(a+h) - \Delta^2 f(a) \\ &= f(a+3h) - 3f(a+2h)] + 3f(a+h)\text{-}f(a) \end{aligned}$$

And so on.

In general the nth, forward difference is given by $\Delta^n f(x) = \Delta^{n-1} f(x+h) - \Delta^{n-1} f(x)$

5.2.2 Backward Differences

The differences f(a+h)-f(a), f(a+2h)-f(a+h), ..., f(a+nh)-f(a+$\overline{n-1}$ h) are called the first difference and these denoted by $\nabla f(a), \nabla f(a+h), \ldots\ldots\ldots, \nabla f(a+nh)$ respectively. But the backward differences are defined by points to the left of x such that

$$\nabla f(a+h) = f(a+h) - f(a)$$
$$\nabla f(a+2h) = f(a+2h) - f(a+h)$$
$$\ldots\ldots\ldots\ldots\ldots\ldots\ldots\ldots\ldots\ldots\ldots\ldots$$
$$\nabla f(a+nh) = f(a+nh) - f(a+\overline{n-1}h)$$

where ∇ is called the **backward differences operator**.

The differences of first backward differences are said to be second backward differences and are denoted by

$$\begin{aligned}\nabla^2 f(a+2h) &= \nabla[f(a+2h) - f(a+h)] \\ &= \nabla f(a+2h) - \nabla f(a+h) \\ &= [f(a+2h) - f(a+h)] - [f(a+h) - f(a)] \\ &= f(a+2h) - 2f(a+h) + f(a)\end{aligned}$$

And so on.

In general first forward difference is given by $\nabla f(x) = f(x) - f(x-h)$

The second backward difference is given by

$$\begin{aligned}\nabla^2 f(x) &= \nabla f(x) - \nabla f(x-h) \\ &= [f(x) - f(x-h)] - [f(x-h) - f(x-2h)] \\ &= f(x) - 2f(x-h)] - -f(x-h) - f(x-2h)]\end{aligned}$$

Proceeding similarly the nth backward differences of y=f(x) is given by

$$\nabla^n f(x) = \nabla[\nabla^{n-1} f(x)] = \nabla^{n-1} f(x) - \nabla^{n-1} f(x-h)$$

5.2.2.1 Properties of operator forward and backward operators

Property 1:

If Δ and ∇ be the first descending difference operator and first ascending difference operator respectively of a function f(x). Show that $(\Delta - \nabla) \equiv \Delta\nabla$.

Solution: By the definition we have

$\Delta f(x) = f(x+h) - f(x)$

and $\nabla f(x) = f(x) - f(x-h)$

Where h is interval of differencing.

$\Delta\nabla f(x) = \Delta[\nabla f(x)] = \Delta[f(x) - f(x-h)]$

$= \Delta f(x) - \Delta f(x-h)$

$= \Delta f(x) - [f(x) - f(x-h)]$

$= \Delta f(x) - \nabla f(x)$

$= (\Delta - \nabla) f(x)$

$\Delta\nabla = (\Delta - \nabla)$

Property 2

The difference of constant function are zero i.e., $\Delta c = 0$, where c is constant.

By definition, $\Delta c = c - c = 0$

Property 3

The operator Δ is (i) cumulative and also (ii) distributive, in its operation as regards constant.

Solution: (i) To prove that $\Delta c = c\Delta$

$(\Delta c)\, f(x) = \Delta c\, f(x) = c\, \{f(x+h) - f(x)\} = c\, \Delta f(x)$

So, $\Delta c = c\Delta$

(ii) To prove that $\Delta[f(x) + g(x)] = [\Delta f(x) + \Delta g(x)]$

$\Delta[f(x) + g(x)] = \{f(x+h) + g(x+h)\} - \{f(x) + g(x)\}$

$= \{f(x+h) - f(x)\} + \{g(x+h) - g(x)\}$

$= \Delta f(x) + \Delta g(x)$

Property 4

The finite difference of a product of two functions is given by

$\Delta[f(x).g(x)] = f(x+h)\, \Delta g(x) + g(x)\, \Delta f(x)$

Solution: $\Delta[f(x).g(x)] = f(x+h)\, g(x+h) - f(x)\, g(x)$

$= f(x+h)\, g(x+h) - f(x+h)\, g(x) + f(x+h)\, g(x) - f(x)\, g(x)$

$= f(x+h)\, [g(x+h) - g(x)] + g(x)\, [f(x+h) - f(x)]$

Therefore, $\Delta[f(x).g(x)] = f(x+h)\, \Delta g(x) + g(x)\, \Delta f(x)$

Property 5

The finite difference of a quotient of two functions is given by

$$\Delta\left[\frac{f(x)}{g(x)}\right] = \frac{g(x)\,\Delta f(x)\ -\ f(x)\,\Delta g(x)}{g(x+h)\; g(x)}$$

Solution: $\Delta\left[\frac{f(x)}{g(x)}\right] = \frac{f(x+h)}{g(x+h)} - \frac{f(x)}{g(x)}$

$$= \frac{f(x+h)\; g(x)\ - f(x)\; g(x+h)}{g(x+h)\; g(x)}$$

$$= \frac{\left[f(x+h)\; g(x)\ - f(x)\; g(x+h)\right] - \left[g(x+h)\; f(x)\ - f(x)\; g(x)\right]}{g(x+h)\; g(x)}$$

$$= \frac{g(x)\ \left[f(x+h)\ - f(x)\right] - f(x)\ \left[g(x+h)\ -\ g(x)\right]}{g(x+h)\; g(x)}$$

Therefore, $\Delta\left[\frac{f(x)}{g(x)}\right] = \frac{g(x)\,\Delta f(x)\ -\ f(x)\,\Delta g(x)}{g(x+h)\; g(x)}$

5.2.3 Central Differences

Let us consider $x_0, x_1, \ldots\ldots, x_n$ be the given set of observations and let $y_0, y_1, \ldots\ldots, y_n$ are the corresponding values of the curve $y = f(x)$, then the central difference operator is denoted by δ, and is defined by

(a) $\delta^n y_{r-\frac{1}{2}} = \delta^{n-1} y_r - \delta^{n-1} y_{r-1}$, where n is odd and $r = 1, 2, 3, \ldots\ldots$

(b) $\delta^n y_r = \delta^{n-1} y_{r+\frac{1}{2}} - \delta^{n-1} y_{r-\frac{1}{2}}$, where n is even and $r = 1, 2, 3, \ldots\ldots$

(c) $\delta^0 y_r = y_r$

The central difference table is shown below

Table 5.1: The standard difference table

x	y	δy	$\delta^2 y$	$\delta^3 y$	$\delta^4 y$
x_0	y_0				
		$\delta y_{\frac{1}{2}}$			
x_1	y_1		$\delta^2 y_1$		
		$\delta y_{\frac{3}{2}}$		$\delta^3 y_{\frac{3}{2}}$	
x_2	y_2		$\delta^2 y_2$		$\delta^4 y_2$
		$\delta y_{\frac{5}{2}}$		$\delta^3 y_{\frac{5}{2}}$	
x_3	y_3		$\delta^2 y_3$		
		$\delta y_{\frac{7}{2}}$			
x_4	y_4				

If h be the common difference in the values of x and y=f(x) be given function then

$\delta f(x) = f(x + h/2) - f(x - h/2)$

5.2.4 Error Propagation in Difference Table

Let $y_0, y_1, \ldots, y_n$ be the true values of a function and suppose the value of y_4 to be affected with an error ε, so that its erroneous value is $y_4 + \varepsilon$. Then the successive difference of y as shown in below table. This table shows that the effect of an error increases with the successive differences, that the coefficients of ε's are the binomial coefficients with alternative signs and the algebraic sum of the errors in any difference column is zero. The same effect is also shown for the horizontal table.

Table 5.2: Error propagation in a difference table

Y	Δy	$\Delta^2 y$	$\Delta^3 y$
y_0			
	Δy_0		
y_1		$\Delta^2 y_0$	
	Δy_1		$\Delta^3 y_0$
y_2		$\Delta^2 y_1$	
	Δy_2		$\Delta^3 y_1$
y_3		$\Delta^2 y_2$	
	Δy_3		$\Delta^3 y_2 + \varepsilon$
y_4		$\Delta^2 y_3 + \varepsilon$	
	$\Delta y_4 + \varepsilon$		$\Delta^3 y_3 - 3\varepsilon$
$y_5 + \varepsilon$		$\Delta^2 y_4 - 2\varepsilon$	
	$\Delta y_5 - \varepsilon$		$\Delta^3 y_4 - 3\varepsilon$
y_6		$\Delta^2 y_5 + \varepsilon$	

Contd.

	Δy_6		$\Delta^3 y_5 + \varepsilon$
y_7		$\Delta^2 y_6$	
	Δy_7		$\Delta^3 y_6$
y_8		$\Delta^2 y_7$	
	Δy_8		
y_9			

Example 5.1: Polynomial of 5th degree is shown in table. It is given that f(3) is in error. Correct the value.

X	0	1	2	3	4	5	6
y = f(x)	1	2	33	254	1054	3126	7777

Solution: It is given that y=f(x) is a polynomial of 5th degree, hence $\Delta^5 y$ must be constant; f(3) is in error.

Let 254+ε be the true value, then construct the difference table.

Table 5.3: Error propagation in a difference table

x	y	Δy	$\Delta^3 y$	$\Delta^3 y$	$\Delta^4 y$	$\Delta^5 y$
0	1					
		1				
1	2		30			
		31		160+ε		
2	33		190+ ε		200-4ε	
		221+ ε		360-3ε		220+10ε
3	254+ε		550-2ε		420+6ε	
		1771 - ε		1780+3ε		20-10ε
4	1054		1330+ε		440-4ε	
		2101		1220-ε		
5	3126		12550			
		4651				
6	7777					

Since the 5th difference of y are constant.

$220 + 10\,\varepsilon = 20 - 10\,\varepsilon$

$20\,\varepsilon = -200$

Or, $\varepsilon = -10$

So, $f(3) = 254 + \varepsilon$

$f(3) = 244$

5.2.5 Other Difference Operator

(a) Shift operator, E: This is another type of difference operator, E, which is given by the rule.

$E\,[f(x)] = f(x + h)$ (1)

or, $E\,y_i = y_{i+1}$

$\Delta\, f(x) = f(x + h) - f(x)$

$\Delta\, f(x) = E\, f(x) - f(x)$ (2)

From equation (2), we have, $E\, f(x) = \Delta f(x) + f(x)$

$\Rightarrow E\, f(x) = (\Delta+1)\, f(x)$

The 2nd shift operator yields

$E^2\, f(x) = E\,[E\, f(x)] = E\,[f(x + h)] = f(x + 2h)$

Therefore, E is linear and obeys the law of indices. In general,

$E^n\, f(x) = f(x + nh)$, or

$E^n\, y_i = y_{i+nh}$

The inverse shift operator, E^{-1}, is: $E^{-1}\, f(x) = f(x - h)$

Similarly, the second order inverse shift operator is, $E^{-2}\, f(x) = f(x - 2h)$

In general, the higher order inverse shift operator is, $E^{-n}\, f(x) = f(x - nh)$

(b) Average operator, μ: This is defined as

$$\mu\, f(x) = \frac{1}{2}\left[f\left(x+\frac{h}{2}\right)-f\left(x-\frac{h}{2}\right)\right]$$

(c) Differential operator, D: This is defined as; $D\, f(x) = \frac{d}{dx} f(x) = f'(x)$

$$D^2\, f(x) = \frac{d^2}{dx^2} f(x) = f''(x)$$

5.2.6 Relationship between the Operators

(1) Operator E and Δ are distributive

Consider a function u(x), which is sum of the functions f(x), g(x) and p(x), ... i.e.,

$u(x) = f(x) + g(x) + p(x) +$ (1)

Then $E\,u(x) = f(x + h) + g(x + h) + p(x + h) + \dots\dots\dots.$, here the interval of differencing is h

$E\,u(x) = E\,f(x) + E\,g(x) + E\,p(x) + \dots\dots\dots.$, i.e., E is distributable.

Again, $\Delta u(x) = \Delta[f(x) + g(x) + p(x) + \dots\dots\dots.]$

$= [f(x+h) + g(x+h) + p(x+h) + \dots\dots\dots.] - [f(x) + g(x) + p(x) + \dots\dots\dots.]$

$= [f(x+h) - f(x)] + [g(x+h) - g(x)] + [p(x+h) - p(x)] \dots\dots\dots.$

$= \Delta f(x) + \Delta g(x) + \Delta p(x) + \dots\dots\dots.$

i.e., Δ is also distributable.

(2) E and Δ are commulative w.r.t constants i.e., $E[c\,f(x)] = c\,E\,f(x)$ and $\Delta[c\,f(x)]=c\,\Delta f(x)$

In the above expression, c is constant.

=> $E[c\,f(x)] = c\,f(x+h) = c\,E\,f(x)$, h being interval of differencing.

Again, $\Delta[c\,f(x)] = c\,f(x+h) - c\,f(x) = c\,\Delta f(x)$

(3) E and Δ obey the law of indices i.e.,

$E^m E^n f(x) = E^{m+n} f(x)$

$\Delta^m \Delta^n f(x) = \Delta^{m+n} f(x)$

=> $E^m E^n f(x) = E^m f(x+nh) = f(x+mh+nh) = f[x+h] = E^{m+n} f(x)$

Similarly, $\Delta^m \Delta^n f(x) = \{\Delta.\,\Delta.\,\Delta \dots\dots\dots. \text{ m times}\}\ \{\Delta.\,\Delta.\,\Delta \dots\dots\dots. \text{ n times}\}\ f(x) = \Delta^{m+n} f(x)$

(4) E and Δ are not commulative w.r.t variables i.e., if $u(x) = f(x).g(x)$

Then $E\,u(x) \neq f(x).E\,g(x)$ and $\Delta u(x) \neq f(x).\,\Delta g(x)$

Similarly, we can prove that

(5) $E\,\Delta = \nabla E = \Delta$

(6) $E^0 f(x) = f(x)$

(7) $E^{-n} f(x) = f(x - nh)$

(8) $E^2 f(x) \neq [E\,f(x)]^2$

(9) If $\Delta f(x) = 0$, then it does not mean that either $\Delta = 0$ or $f(x) = 0$

(10) $(1+\Delta)(1-\nabla) = 1$

=> $(1+\Delta)(1-\nabla) f(x) = (1+\Delta)\{(1-\nabla) f(x)\}$

$= (1+\Delta)\ \{f(x) - f(x)\}$

$= (1+\Delta)\ [f(x) - \{f(x) - f(x-h)\}]$

$= (1+\Delta)\ f(x-h)$

$= E\ f(x-h)$ [Because $E = 1 + \Delta$]

$= f(x)$

$= 1.f(x)$

Therefore, $(1+\Delta)\ (1-\nabla)\ f(x) = 1.f(x), \forall f(x)$.

Hence proved, $(1+\Delta)\ (1-\nabla) = 1$

5.2.7 Representation of a Polynomial using Factorial Notation

The product factors in which the 1st factor is x and the successive factors decrease by a constant difference is known as a factorial function and is denoted by $x^{(n)}$, where n is a positive integer. These functions play an important role in the theory of finite differences with the help of these functions. We can find the various order differences directly and by simple rule of differentiation.

When the interval of differentiating is unity;

$$x^{(n)} = x(x-1)(x-2) \ldots\ldots\ldots\ldots (x-\overline{n-1}) \quad (1)$$

When the interval of differentiation is h;

$$x^{(n)} = x(x-h)(x-2h) \ldots\ldots\ldots\ldots (x-\overline{n-1}\,h) \quad (2)$$

Again from equation (1)

$$x^{(n)} = \frac{x(x-1)(x-2) \ldots\ldots\ldots\ldots (x-\overline{n-1})\ (x-n)!}{(x-n)!}$$

$$x^{(n)} = \frac{x!}{(x-n)!};(x > n)$$

Theorem 1 $\Delta^n x^{(n)} = n!h^n$ and $\Delta^{n+1} x^{(n)} = 0$, where Δ is an operator

Since

$$\Delta x^{(n)} = (x+h)^{(n)} - x^{(n)}$$

$$\Delta x^{(n)} = (x+h)(x+h-h)(x+h-2h) \ldots\ldots\ldots\ldots (x+h-\overline{n-1}h) - [x(x-h) \ldots\ldots\ldots\ldots (x-\overline{n-1}h)]$$

$$= (x+h)(x)(x-h) \ldots\ldots\ldots\ldots (x-\overline{n-2}h) - x(x-h) \ldots\ldots\ldots\ldots (x-\overline{n-2}h)(x-\overline{n-1}h)$$

$$= x(x-h) \ldots\ldots\ldots\ldots (x-\overline{n-2}h)[(x+h)-(x-\overline{n-1}h)]$$

$$= x^{(n-1)}[x+h-x+nh-h]$$

Table 5.4: Definition and relationship between operators

Operator	Formula	Conversion into			
		E	Δ	∇	δ
Δ, Forward difference operator	$\Delta f(x) = f(x+h) - f(x)$	$E-1$		$(1-\nabla)^{-1}-1$	$\frac{1}{2}\delta^2 + \delta\sqrt{\left(1+\frac{1}{4}\delta^2\right)}$
∇, Backward difference operator	$\nabla f(x) = f(x) - f(x-h)$	$1-E^{-1}$	$1-(1+\Delta)^{-1}$	∇	$-\frac{1}{2}\delta^2 + \delta\sqrt{\left(1+\frac{1}{4}\delta^2\right)}$
δ, Central difference operator	$\delta f(x) = f\left(x+\frac{h}{2}\right) - f\left(x-\frac{h}{2}\right)$	$E^{1/2}-E^{-1/2}$	$\Delta(1+\Delta)^{-1/2}$	$\nabla(1-\nabla)^{-1/2}$	δ
E, Shift operator	$E f(x) = f(x+h)$	E	$\Delta+1$	$(1-\nabla)^{-1}$	$1+\frac{1}{2}\delta^2 + \delta\sqrt{\left(1+\frac{1}{4}\delta^2\right)}$
μ, Average operator	$\mu f(x) = \frac{1}{2}\left[f\left(x+\frac{h}{2}\right) - f\left(x-\frac{h}{2}\right)\right]$	$0.5(E^{1/2}+E^{-1/2})$	$\left(1+\frac{\Delta}{2}\right)(1+\Delta)^{1/2}$	$\left(1-\frac{\Delta}{2}\right)(1-\Delta)^{-1/2}$	$\sqrt{\left(1+\frac{1}{4}\delta^2\right)}$
D, Differential operator	$D f(x) = f'(x)$	-	-	-	-

$$\Delta x^{(n)} = nhx^{(n-1)} \tag{3}$$

Equivalently, $\dfrac{\Delta x^{(n)}}{\Delta x} = nx^{(n-1)}$ (Because $\Delta x = h$)

Again, from equation (3),

$$\begin{aligned}
\Delta^2 x^{(n)} &= \Delta.\Delta x^{(n)} = nh\Delta x^{(n-1)} \\
&= nh\Delta[x(x-h)...........(x-\overline{n-2}h) \\
&= nh[(x+h)(x)(x-h)...........(x+h-\overline{n-2}h)-[x(x-h)...........(x-\overline{n-2}h)] \\
&= nh[x(x-h)...........(x-\overline{n-3}h)(x+h-x+\overline{n-2}h)] \\
&= n(n-1)h^2 x(x-h)...........(x-\overline{n-3}h) \\
&= n(n-1)h^2 x^{(n-2)}
\end{aligned}$$

Proceeding similarly, we get, $\Delta^{n-1} x^{(n)} = n(n-1)...........2.1.h^{n-1}x$

Therefore,

$$\begin{aligned}
\Delta^n x^{(n)} &= n(n-1)...........2.1.h^{n-1}\Delta x \\
&= n(n-1)...........2.1.h^{n-1}(x+h-x) \\
&= n(n-1)...........2.1.h^n
\end{aligned}$$

Hence, $\Delta^n x^{(n)} = n!h^n$ (4)

Again, $\Delta^{n+1} x^{(n)} = \Delta(n!h^n) = n!h^n - n!h^n = 0$

In particular, when $h = 1$, we get from equation (4) that, $\Delta^n x^{(n)} = n!$

Theorem 2: To show that $x^{(-n)} = \dfrac{1}{(x+n)^{(n)}}$, the interval of differentiating being unity.

From the definition of $x^{(n)}$, when then interval of differentiating is h, we have

$$x^{(n)} = (x-\overline{n-1}h)x^{(n-1)} \tag{5}$$

Therefore, when $n = 0$, we have, $x^{(0)} = (x+h)x^{(-1)}$

Again, $x^{(0)} = 1$.

[Since, $\Delta x^{(n)} = n\,h\,x^{(n-1)}$

Therefore when $n = 1$; $\Delta x^{(1)} = h\,x^{(0)}$

Again $\Delta x^{(1)} = \Delta x = h; \Rightarrow x^{(0)} = 1$]

Therefore, $x^{(-1)} = \dfrac{1}{x+h}$

When n = –1, then from equation (5), $x^{(-1)} = (x + 2h)\, x^{(-2)}$

$$\frac{1}{x+h} = (x+2h)x^{(-2)}$$

$$\Rightarrow x^{(-2)} = \frac{1}{(x+n)(x+h)}$$

In general, $$x^{(-n)} = \frac{1}{(x+h)(x+2h)..........(x+nh)} \quad (6)$$

$x^{(-n)} = \dfrac{1}{(x+nh)^{(n)}}$, where $x^{(-n)}$ is known as reciprocal factorial and n is a positive integer.

Particular case: If h = 1, then

$$x^{(-n)} = \frac{1}{(x+1)(x+2)(x+3)..........(x+n)} = \frac{1}{(x+n)^{(n)}}$$

Example 5.2: If m is any integer and c is any constant, show that

(a) $\dfrac{\Delta}{\Delta x}\left[cx^{(n)}\right] = c\dfrac{\Delta}{\Delta x}x^{(n)} = ncx^{(n-1)}$

(b) $\Delta\left[cx^{(n)}\right] = c\Delta x^{(n)} = nchx^{(n-1)}$

Since Δ and $\dfrac{\Delta}{\Delta x}$ are linear operations and we have

$\dfrac{\Delta}{\Delta x}\left[cx^{(n)}\right] = c\dfrac{\Delta}{\Delta x}x^{(n)}$ and $\Delta\left[cx^{(n)}\right] = c\Delta x^{(n)}$

[The results are derived by Theoren `1 and Theorem 2

Example 5.3: Find values of

(a) $\dfrac{\Delta}{\Delta x}\left[2x^{(5)}\right]$ (b) $\Delta\left[9x^{(4)}\right]$ (c) $\dfrac{\Delta}{\Delta x}\left[3x^{(-2)}\right]$ (d) $\Delta\left[-5x^{(-4)}\right]$

The interval of differencing is h.

Solution:

(a) $\frac{\Delta}{\Delta x}\left[2x^{(5)}\right] = 5.2.x^{(4)} = 10x^{(4)} = 10x(x-h)(x-2h)(x-3h)$

(b) $\Delta\left[9x^{(4)}\right] = 4.9.x^{(3)} = 36hx^{(4)} = 36hx(x-h)(x-2h)$

(c) $\frac{\Delta}{\Delta x}\left[3x^{(-2)}\right] = (-4)(-5)h.x^{(-3)} = -6x^{(-3)} = \frac{-6}{(x+h)(x+2h)(x+3h)}$

(d) $\Delta\left[-5x^{(-4)}\right] = (-4)(-5).h.x^{(-5)} = 20.h.x^{(-5)} = \frac{20h}{(x+h)(x+2h)(x+3h)(x+4h)(x+5h)}$

Example 5.4: Express $f(x) = 3x^3 + x^2 + x + 1$, in the factorial notation and the differencing interval being unity.

Solution: Here f(x) is a polynomial of 3rd degree, so we can write

$$f(x) = f(0) + \frac{\Delta f(0)}{1!}x^1 + \frac{\Delta^2 f(0)}{2!}x^2 + \frac{\Delta^3 f(0)}{3!}x^3$$

The differencing interval is 1 and hence finding the function values at x = 0, 1, 2, 3, we get

f(0) = 1, f(1) = 6, f(2) = 31, f(3) = 94.

The difference table is

x	f(x)	Δf(x)	$\Delta^2 f(x)$	$\Delta^3 f(x)$
0	1			
		5		
1	6		20	
		25		16
2	31		38	
		62		
3	94			

Through table, the values are f(0) = 1, Δf(0) = 5, $\Delta^2 f(0) = 20$, $\Delta^3 f(0) = 18$ and put in f(x),

$$f(x) = 1 + 5x^1 + \frac{20}{2!}x^2 + \frac{18}{3!}x^3$$

Hence $f(x) = 3x^3 + 10x^2 + 5x + 1$

Example 5.5: Evaluate (a) $\Delta^2 e^x$ (b) $\Delta \tan^{-1} x$

Solution:

(a) $\Delta^2 e^x = \Delta(\Delta e^x) = \Delta[e^{x+h} - e^x] = \Delta[e^x (e^h - 1)] = (e^h - 1)\, \Delta e^x = (e^h - 1)(e^{x+h} - e^x) = (e^h - 1)\, e^x$

Hence $\Delta^2 e^x = (e^h - 1)\, e^x$

(b) $\Delta \tan^{-1} x = \tan^{-1}(x+h) - \tan^{-1} x = \tan^{-1}\left[\dfrac{x+h-x}{(1+(x+h)x}\right] = \tan^{-1}\left[\dfrac{h}{1+hx+x^2}\right]$

5.3 Interpolation with Equal Intervals

Interpolation is a method to estimate the new data points (or unknown intermediate values) of a function [i.e., y = f(x)] within the range of a discrete set of known independent data point/variable i.e., x. **Extrapolation** is a method to estimate value based on extending a known sequence of values or facts beyond the area that is certainly known. Extrapolation is included in Interpolation. Interpolation is based on the calculus of finite differences. Consider, equation is $y = x^2$, where function is $f(x) = x^2$; then the value of y for a set of values of x is

x:	1	3	5
y:	1	9	25

Interpolation is finding the function values when x ranges between 1 to 5, and extrapolation finding function value outside the range of x.

If the function f(x) is explicitly known, then value of y can be easily found. But if the function value is not known then it is difficult to find the exact form of function. So to find the solution, the function is replaced by a simpler function $\phi(x)$, called interpolating function (or smoothing function). If $\phi(x)$ is polynomial, then it is called the interpolating polynomial and the process is called the polynomial interpolation. Similarly, for trigonometric series the function is called trigonometric interpolation.

5.3.1 Newton's Forward Interpolation Formula

This formula is used to interpolate values of y near the beginning of a set of tabular values. Suppose, function y = f(x), take the values $y_0, y_1, y_2, \ldots\ldots\ldots y_n$ corresponding to the values $x_0, x_1, x_2, \ldots\ldots\ldots x_n$.

Suppose, $x_i = x_0 + ih$ (1)

where $i = 0, 1, 2, \ldots\ldots\ldots, n$; and h is differencing interval.

Suppose $\phi(x)$ is the polynomial of nth degree in x such that;

$\phi(x_0) = y_0,\ \phi(x_1) = y_1,\ \phi(x_2) = y_2,\ \ldots\ldots\ldots,\ \phi(x_n) = y_n$ (2)

It can be written as equation (3) below

$\phi(x) = a_0 + a_1 (x - x_0) + a_2(x - x_0)(x - x_1) + \ldots\ldots\ldots + a_n (x - x_0)(x - x_1) \ldots\ldots\ldots (x - x_{n-1})$

where a_0, a_1, a_2, ………, a_n are constants, which have to be determined.

Put $x = x_0$ in equation (3), which yields: $\phi(x_0) = a_0$;

and on comparing with equation (2): $y_0 = a_0$

Put $x = x_1$ in equation (3), which yields: $\phi(x_1) = a_0 + a_1 (x_1 - x_0)$

$y_1 = y_0 + a_1 (x_1 - x_0)$

$$a_1 = \frac{y_1 - y_0}{x_1 - x_0} = \frac{\Delta y_0}{h}$$

Where $h = x_1 - x_0$, is obtained from equation (1).

Put $x = x_2$ in equation (3), which yields: $\phi(x_2) = a_0 + a_1 (x_2 - x_0) + a_2(x_2 - x_0)(x_2 - x_1)$

$y_2 = a_0 + a_1 (x_2 - x_0) + a_2(x_2 - x_0)(x_2 - x_1)$

$$y_2 = y_0 + \frac{\Delta y_0}{h}(2h) + a_2(2h)(h); \text{ where } x_2 = x_0 + 2h$$

$$a_2 = \frac{\Delta y_1 - \Delta y_0}{2h^2} = \frac{\Delta^2 y_0}{2!\ h^2}$$

Similarly $a_3 = \dfrac{\Delta^3 y_0}{3!\ h^3}$, ………, $a_n = \dfrac{\Delta^n y_0}{n!\ h^n}$

Put the values of a_0, a_1, a_2, ………, a_n in equation (3), yield equation (4) as below

$$\phi(x) = y_0 + \frac{\Delta y_0}{h}(x - x_0) + \frac{\Delta^2 y_0}{2!\ h^2}(x - x_0)(x - x_1) + \ldots\ldots\ldots + \frac{\Delta^n y_0}{n!\ h^n}(x - x_0)(x - x_1)\ldots(x - x_{n-1})$$

Suppose $p = \dfrac{x - x_0}{h}$, or $x - x_0 = ph$

So, $x - x_1 = x - x_0 + x_0 - x_1 = (x - x_0) - (x_1 - x_0) = ph - h = (p - 1)h$

$x - x_2 = x - x_1 + x_1 - x_2 = (x - x_1) - (x_2 - x_1) = (p - 1)h - h = (p - 2)h$

Similarly, $x - x_3 = (p - 3)h$

……….

$x - x_{n-1} = (p + n - 1)\,h$

Put all these values in equation (4), which yields equation (5) as

$$\phi(x) = y_0 + \frac{p}{1!}\Delta y_0 + \frac{p\,(p-1)}{2!}\Delta^2 y_0 + \ldots + \frac{p\,(p-1)\,\ldots\ldots\,(p-n+1)}{n!}\Delta^n y_0 \quad (5)$$

Equation (5) is called Newton's forward difference interpolation formula or Newton-Gregory formula for forward interpolation with equal intervals.

$$\phi(x) = y_0 + \frac{p^{(1)}}{1!}\Delta y_0 + \frac{p^{(2)}}{2!}\Delta^2 y_0 + \ldots\ldots + \frac{p^{(n)}}{n!}\Delta^n y_0 \quad (6)$$

where $p^{(2)} = p\,(p-1)$, ..., $p^{(n)} = p\,(p-1)\,\ldots\ldots\,(p-n+1)$

Example 5.6: The population of town in the decennial census was given below. Estimate the population for the year 1895.

Year (x)	:	1891	1901	1911	1921	1931
Population (y) (in thousands)	:	46	66	81	93	101

Solution: The difference table is

Year (x)	Population (y)	Δy	$\Delta^2 y$	$\Delta^3 y$	$\Delta^4 y$
1891	46				
		20			
1901	66		–5		
		15		2	
1911	81		–3		–3
		12		–1	
1921	93		–4		
		8			
1931	101				

x_0 (previous year) = 1891, x (population of year to be find) = 1895, and

h (difference in year in table) = 1901 – 1891 = 1911 – 1901 = 10

$$p = \frac{x - x_0}{h} = \frac{1895 - 1891}{10} = 0.4$$

Newton's forward difference interpolation formula is

$$\phi(x) = y_0$$

$$+ \frac{p}{1!}\Delta y_0 + \frac{p\,(p-1)}{2!}\Delta^2 y_0 + \frac{p\,(p-1)\,(p-2)}{3!}\Delta^3 y_0 + \frac{p\,(p-1)\,(p-2)\,(p-3)}{4!}\Delta 4 y_0$$

$$x(1895) = 46 + (0.4 * 20) + \frac{(0.4)(0.4 - 1)}{2}(-5) + \frac{(0.4)(0.4 - 1)(0.4 - 2)}{3 * 2 * 1}(2) + \frac{(0.4)(0.4 - 1)(0.4 - 2)(0.4 - 3)}{4 * 3 * 2 * 1}(-3)$$

x(1895) = 54.85 thousands

Example 5.7: Find the number of men getting wages between 10 Rs and 15 Rs from the following data:

Wages (in Rs)	0-10	10-20	20-30	30-40
Men (in number)	9	30	35	42

Solution: The difference table is

Wages (x)	Men (y)	Δy	$\Delta^2 y$	$\Delta^3 y$
<10	9			
		30		
<20	9 + 30 =39		5	
		35		2
<30	39 + 35 = 74		7	
		42		
<40	74 + 42 =116			

For finding the number of men getting wages between 10-15 Rs,

so x_0 (initial wage) = 10, x (final wage) = 15

h (difference in wage) = 10

$$p = \frac{x - x_0}{h} = \frac{15-10}{10} = 0.5$$

Newton's forward difference interpolation formula is

$$\phi(x) = y_0 + \frac{p}{1!}\Delta y_0 + \frac{p(p-1)}{2!}\Delta^2 y_0 + \frac{p(p-1)(p-2)}{3!}\Delta^3 y_0 + \ldots..$$

$$\phi(15) = 9 + (0.5 * 30) + \frac{(0.5)(0.5 - 1)}{2}(5) + \frac{(0.5)(0.5 - 1)(0.5 - 2)}{3 * 2 * 1}(2)$$

$\phi(15) = 23.5 = 24$ nearly

Hence number of men getting wages between 10 Rs and 15 Rs = 24 – 9 = 15

5.3.2 Newton's Backward Interpolation Formula

This formula is used to interpolate the values of $y = \phi(x)$ near the end of the set of given values. Suppose, function $y = f(x)$, take the values $y_0, y_1, y_2, \ldots\ldots\ldots y_n$ corresponding to the values $x_0, x_1, x_2, \ldots\ldots\ldots x_n.$

Suppose, $x_i = x_0 + ih$ (1)

where $i = 0, 1, 2, \ldots\ldots\ldots, n$; and h is differencing interval.

Suppose $\phi(x)$ is the polynomial of nth degree in x such that;

$$\phi(x_0) = y_0,\ \phi(x_1) = y_1,\ \phi(x_2) = y_2,\ \ldots\ldots\ldots,\ \phi(x_n) = y_n \quad (2)$$

It can be written as equation (3) below

$$\phi(x) = a_0 + a_1 (x - x_n) + a_2(x - x_n)(x - x_{n-1}) + \ldots\ldots\ldots + a_n (x-x_n)(x-x_{n-1}) \ldots\ldots(x-x_2)(x-x_1) \quad (3)$$

where $a_0, a_1, a_2, \ldots\ldots\ldots, a_n$ are constants, which have to be determined.

Put $x = x_n$ in equation (3), which yields: $\phi(x_n) = a_0$;

and on comparing with equation (2): $y_n = a_0$

Put $x = x_{n-1}$ in equation (3), which yields: $\phi(x_{n-1}) = a_0 + a_1 (x_{n-1} - x_n)$

$$y_{n-1} = y_n + a_1 (x_{n-1} - x_n)$$

$$a_1 = \frac{y_n - y_{n-1}}{x_n - x_{n-1}} = \frac{\nabla y_n}{h}$$

Where $h = x_1 - x_0$, is obtained from equation (1).

Put $x = x_{n-2}$ in equation (3), which yields:

$$\phi(x_{n-2}) = a_0 + a_1 (x_{n-2} - x_n) + a_2(x_{n-2} - x_n)(x_{n-2} - x_{n-1})$$

$$y_{n-2} = y_n + \frac{\nabla y_n}{h}(-2h) + a_2(-2h)(-h)$$

$$a_2 = \frac{\nabla y_n - \nabla y_{n-1}}{2h^2} = \frac{\nabla^2 y_n}{2!\ h^2}$$

Similarly, $a_2 = \dfrac{\nabla y_n - \nabla y_{n-1}}{2h^2} = \dfrac{\nabla^2 y_n}{2!\ h^2}$

Put the values of $a_0, a_1, a_2, \ldots\ldots\ldots, a_n$ in equation (3), yield equation (4) as below

$\phi(x) = y_n + \frac{\nabla y_n}{h}(x - x_n) + \frac{\nabla^2 y_n}{2!\ h^2}(x - x_n)(x - x_{n-1}) + \ldots\ldots\ldots + \frac{\nabla^n y_n}{n!\ h^n}(x - x_n)(x - x_{n-1})\ldots(x - x_1)$ (4)

Suppose $p = \frac{x - x_n}{h}$, or $x - x_n = ph$

So, $x - x_{n-1} = x - x_n + x_n - x_{n-1} = (x - x_n) + (x_n - x_{n-1}) = ph + h = (p + 1) h$

$x - x_{n-2} = x - x_{n-1} + x_{n-1} - x_{n-2} = (p + 1) h + h = (p + 2) h$

Similarly, $x - x_{n-3} = (p + 3) h$

....

$x - x_1 = (p + n - 1) h$

Put all these values in equation (4), which yields equation (5) as

$\phi(x) = y_n + \frac{p}{1!}\nabla y_n + \frac{p(p+1)}{2!}\nabla^2 y_n + \ldots + \frac{p(p+1)\ldots\ldots(p+n-1)}{n!}\nabla^n y_n + \ldots$ (5)

Equation (5) is called Newton's backward difference interpolation formula or Newton-Gregory formula for backward interpolation with equal intervals.

$\phi(x) = y_0 + \frac{p^{(1)}}{1!}\Delta y_0 + \frac{p^{(1)}}{1!}\Delta y_0 + \ldots\ldots + \frac{p^{(n)}}{n!}\Delta^n y_0$ (6)

where $p^{(2)} = p(p - 1)$, ..., $p^{(n)} = p(p - 1)\ \ldots\ldots\ldots\ (p - n + 1)$

Example 5.8: The following data gives the melting point of an alloy of lead and zinc, where t is the temperature in degrees and P is the percentage of lead in the alloy.

P	40	50	60	70	80	90
T	180	204	226	250	276	304

Find the melting point of the alloy containing 84% lead.

Solution: This value lies near the end of table, so the difference table is

P	T	∇y	$\nabla^2 y$	$\nabla^3 y$	$\nabla^4 y$	$\nabla^5 y$
40	180					
		20				
50	204		2			
		22		0		
60	226		2		0	

Contd.

		24		0		0
70	250		2		0	
		26		0		
80	276		2			
		28				
90	304					

$x_n = 90$, x (to find t at lead percentage) = 84, h (difference in P) = 10

$$p = \frac{x - x_n}{h} = \frac{84 - 90}{10} = -0.6$$

Newton's backward interpolation formula is

$$\phi(x) = y_n + \frac{p}{1!}\nabla y_n + \frac{p\,(p+1)}{2!}\nabla^2 y_n + ...$$

$$\phi(84) = 304 - (0.6 * 28) + \frac{(-0.6)\,(-0.6 + 1)}{2 * 1}\,(2)$$

$$\phi(84) = 286.96$$

5.4 Interpolation with Unequal Intervals

The above section 5.3 deals with interpolation with equal intervals and the method fails for unequal distribution of values. So to deal with unequal intervals, the methods are

(a) Lagrange's interpolation formula

(b) Newton's general interpolation formula with divided differences

5.4.1 Lagrange's Interpolation Formula

Suppose $y = f(x)$ is the function, which takes the values (x_0, y_0), (x_1, y_1) .... (x_n, y_n). Since there are (n+1) pairs of values of x and y, so function f(x) is a polynomial of x of degree n as

$$y = f(x) = a_0\,(x - x_1)\,(x - x_2)\,\ldots\,(x - x_n) + a_1\,(x - x_0)\,(x - x_2)\,\ldots\,(x - x_n) + \ldots\ldots\ldots$$

$$+ a_n\,(x - x_0)\,(x - x_1)\,\ldots\,(x - x_{n-1}) \qquad (1)$$

Put $x = x_0$, $y = y_0$ in above equation, yielding; $a_0 = \dfrac{y_0}{(x_0 - x_1)\,(x_0 - x_1)\,...\,(x_0 - x_n)}$

On putting $x = x_1$, $y = y_1$ in equation (1), yielding;

$$a_1 = \frac{y_1}{(x_1 - x_0)\,(x_1 - x_2)\,...\,(x_1 - x_n)}$$

On putting $x = x_n$, $y = y_n$ in equation (1), yielding;

$$a_n = \frac{y_n}{(x_n - x_0)(x_n - x_1)\dots(x_n - x_{n-1})}$$

Put above three equations in equation (1) as

$$f(x) = \frac{(x - x_1)(x - x_2)\dots(x - x_n)}{(x_0 - x_1)(x_0 - x_2)\dots(x_0 - x_n)} y_0 + \frac{(x - x_0)(x - x_2)\dots(x - x_n)}{(x_1 - x_0)(x_1 - x_2)\dots(x_1 - x_n)} y_1 + \dots +$$

$$\frac{(x - x_0)(x - x_1)\dots(x - x_{n-1})}{(x_n - x_0)(x_n - x_1)\dots(x_n - x_{n-1})} y_n$$

This above equation is called Lagrange's interpolation formula. The disadvantage of this method is recalculation of interpolation coefficients on insertion of additional value and process become slow for solving higher degree polynomial because of large number of calculations. This limitation is removed by Newton's general interpolation formula with divided differences.

Example 5.9: Using Lagrange's interpolation formula, find the value of y corresponding to x = 10 from the following table:

x	5	6	9	11
y	12	13	14	16

Solution: From the table, $x_0 = 5, x_1 = 6, x_2 = 9, x_3 = 11$ and

$y_0 = 12, y_1 = 13, y_2 = 14, y_3 = 16$

Using Lagrange's interpolation formula

$$f(x) = \frac{(x - x_1)(x - x_2)(x - x_3)}{(x_0 - x_1)(x_0 - x_2)(x_0 - x_3)} y_0 + \frac{(x - x_0)(x - x_2)(x - x_3)}{(x_1 - x_0)(x_1 - x_2)(x_1 - x_3)} y_1$$

$$\frac{(x - x_0)(x - x_1)(x - x_3)}{(x_2 - x_0)(x_2 - x_1)(x_2 - x_3)} y_2 + \frac{(x - x_0)(x - x_1)(x - x_2)}{(x_3 - x_0)(x_3 - x_1)(x_3 - x_2)} y_3$$

Put all the values

$$f(x) = \frac{(10 - 6)(10 - 11)(10 - 11)}{(5 - 6)(5 - 9)(5 - 11)} \times 12 + \frac{(10 - 5)(10 - 9)(10 - 11)}{(6 - 5)(6 - 9)(6 - 11)} \times 13$$

$$\frac{(10 - 5)(10 - 6)(10 - 11)}{(9 - 5)(9 - 6)(9 - 11)} \times 14 + \frac{(10 - 5)(10 - 6)(10 - 9)}{(11 - 5)(11 - 6)(11 - 9)} \times 16$$

$$f(x) = \frac{44}{3} = 14.6667$$

5.4.2 Newton's General Interpolation Formula with Divided Differences

Suppose the given points are (x_0, y_0), (x_1, y_1), (x_2, y_2) .... $(x_{\backslash n}, y_n)$.

The 1st divided difference for values x_0 and x_1; $[x_0, x_1]$ $\frac{(y_1 - y_0)}{(x_1 - x_0)} = \frac{(y_0 - y_1)}{(x_0 - x_1)}$

Similarly 1st divided difference for values x_1 and x_2; $[x_1, x_2] = \frac{(y_2 - y_1)}{(x_2 - x_1)}$

and so on.

The 2nd divided difference for values x_0, x_1 and x_2; $[x_0, x_1, x_2]$

$$= \frac{[x_1, x_2] - [x_0, x_1]}{x_2 - x_0}$$

$$= \frac{[x_0, x_1] - [x_1, x_2]}{x_0 - x_2}$$

Similarly the 2nd divided difference for values x_1, x_2 and x_3; $[x_1, x_2, x_3] =$

and so on.

The 3rd divided difference for values x_0, x_1, x_2, x_3; $[x_0, x_1, x_2, x_3]$

$$\frac{[x_0, x_1] - [x_1, x_2]}{x_0 - x_2} = \frac{[x_0, x_1, x_2] - [x_1, x_2, x_3]}{x_0 - x_3}$$

Similarly the 3rd divided difference for values x_1, x_2, x_3, x_4;

$$[x_1, x_2, x_3, x_4] = \frac{[x_2, x_3, x_4] - [x_1, x_2, x_3]}{x_4 - x_0}$$

and so on.

The divided differences are symmetrical in their values and independent of the order of the arguments. The nth divided differences of a polynomial of the nth degree are constant.

Table 5.5: The divided difference

X	Y	1st divided difference	2nd divided difference	3rd divided difference
x0	y0			
		$\frac{y_1 - y_0}{x_1 - x_0}$ = a1 (suppose)		
x1	y1		$\frac{a_2 - a_1}{x_2 - x_0}$ = b1	
		$\frac{y_2 - y_1}{x_2 - x_1}$ = a2		$\frac{b_2 - b_1}{x_3 - x_0}$ = c1
x2	y2		$\frac{a_3 - a_2}{x_3 - x_1}$ = b2	
		$\frac{a_3 - a_2}{x_3 - x_1}$ = a3		
x3	y3			

Suppose function $y = f(x)$ and $y_0, y_1 \ldots. y_n$ are the values of arguments $x_0, x_1, \ldots x_n$ respectively.

For 1st divided difference; $[x, x_0] = \frac{(y - y_0)}{(x - x_0)} \Rightarrow y = y_0 + (x - x_0)[x, x_0]$ (2)

For 2nd divided difference;

$[x, x_0, x_1] = \frac{[x, x_0] - [x_0, x_1]}{x - x_1} \Rightarrow [x, x_0] = [x_0, x_1] + (x - x_0)[x, x_0, x_1]$

Put value of $[x, x_0]$ in equation (2), yielding

$$y = y_0 + (x - x_0)[x_0, x_1] + (x - x_0)(x - x_1)[x, x_0, x_1] \quad (3)$$

For 3rd divided difference; $[x, x_0, x_1, x_2] = \frac{[x, x_0, x_1] - [x_0, x_1, x_2]}{x - x_2}$

$\Rightarrow [x, x_0, x_1] = [x_0, x_1, x_2] + (x - x_2)[x, x_0, x_1, x_2]$

Put value of $[x, x_0, x_1]$ in equation (3), yielding

$$y = y_0 + (x - x_0)[x_0, x_1] + (x - x_0)(x - x_1)[x_0, x_1, x_2] + (x - x_0)(x - x_1)(x - x_2)[x, x_0, x_1, x_2]$$

On proceeding with same way, the resulting equation is

$$f(x) = y = y_0 + (x - x_0)[x_0, x_1] + (x - x_0)(x - x_1)[x_0, x_1, x_2] + (x - x_0)(x - x_1)(x - x_2)[x_0, x_1, x_2, x_3] + \ldots + (x - x_0)(x - x_1) \ldots (x - x_n)[x, x_0, \ldots x_n]$$

This above equation is called Newton's general interpolation formula with divided differences.

Note: Relationship between divided differences and forward differences: Consider $(x_0, y_0), (x_1, y_1), (x_2, y_2), \ldots (x_n, y_n)$ are the values. $x_{0,} x_1, \ldots x_n$ are 'h' equispaced. $\Delta y_0 = y_1 - y_0$

$$[x_0, x_1, \ldots x_n] = \frac{\Delta^n y_0}{n!\, h^n}$$

Example 5.10: Given the values of x and f(x) as follows

x	5	7	11	13	17
y	150	392	1452	2366	5202

Find f(x) at x = 9 using Newton's divided difference formula.

Solution: The divided difference table is

X	y = f(x)	1st divided difference	2nd divided difference	3rd divided difference	4th divided difference
$x_0 = 5$	$y_0 = 150$	$\frac{392-150}{7-5} = 121$			
$x_1 = 7$	$y_1 = 392$		$\frac{265-121}{11-5} = 24$		
		$\frac{1452-392}{11-7} = 265$		$\frac{32-24}{13-5} = 1$	
$x_2 = 11$	$y_2 = 1452$		$\frac{457-265}{13-7} = 32$		$\frac{1-1}{17-5} = 0$
		$\frac{2366-1452}{13-11} = 457$		$\frac{42-32}{17-7} = 1$	
$x_3 = 13$	$y_3 = 2366$		$\frac{709-457}{17-11} = 42$		
		$\frac{5302-2366}{17-13} = 709$			
$x_4 = 17$	$y_4 = 5302$				

By Newton's divided difference formula

$f(x) = y_0 + (x - x_0)\,[x_0, x_1] + (x - x_0)(x - x_1)\,[x_0, x_1, x_2] + (x - x_0)(x - x_1)(x - x_2)\,[x_0, x_1, x_2, x_3]$

$f(x) = 150 + (9 - 5)\,121 + (9 - 5)\,(9 - 7)\,24 + (9 - 5)\,(9 - 7)\,(9 - 11)\,1 = 810$

5.5 Central Difference Interpolation Formula

This method is used for interpolation near the middle values of a tabulated data. Consider, the function $y = f(x)$. The values of y are $y_{-2}, y_{-1}, y_0, y_1, y_2$ for the values of x as x_0-2h, x_0-h, x_0, x_0+h, x_0+2h respectively. The spacing between the two consecutive values of x is same. Supoose, $y = y_0$ represents the central ordinate corresponding to $x = x_0$. The difference table is

X	Y	1st difference, Δy	2nd difference, $\Delta^2 y$	3rd difference, $\Delta^3 y$	4th difference, $\Delta^4 y$	5th difference, $\Delta^5 y$	6th difference, $\Delta^6 y$
x_0-3h	y_{-3}						
		Δy_{-3} $(= \delta_{-5/2})$					
x_0-2h	y_{-2}		$\Delta^2 y_{-3}$ $(= \delta^2_{-2})$				
		Δy_{-2} $(= \delta_{-3/2})$		$\Delta^3 y_{-3}$ $(= \delta^3_{-3/2})$			
x_0-h	y_{-1}		$\Delta^2 y_{-2}$ $(= \delta^2_{-1})$		$\Delta^4 y_{-3}$ $(= \delta^4_{-1})$		
		Δy_{-1} $(= \delta_{-1/2})$		$\Delta^3 y_{-2}$ $(= \delta^3_{-1/2})$		$\Delta^5 y_{-3}$ $(= \delta^5_{-1/2})$	
x_0	y_0		$\Delta^2 y_{-1}$ $(= \delta^2_0)$		$\Delta^4 y_{-2}$ $(= \delta^4_0)$		$\Delta^6 y_{-3}$ $(= \delta^6_0)$
		Δy_0 $(= \delta_{1/2})$		$\Delta^3 y_{-1}$ $(= \delta^3_{1/2})$		$\Delta^5 y_{-2}$ $(= \delta^5_{1/2})$	
x_0+h	y_1		$\Delta^2 y_0$ $(= \delta^2_1)$		$\Delta^4 y_{-1}$ $(= \delta^4_1)$		
		Δy_1 $(= \delta_{3/2})$		$\Delta^3 y_0$ $(= \delta^3_{3/2})$			
x_0+2h	y_2		$\Delta^2 y_1$ $(= \delta^2_2)$				
		Δy_2 $(= \delta_{5/2})$					
x_0+3h	y_3						

Conclusion from the above table is

$\Delta y_0 = y_1 - y_0$ or $y_0 = y_1 + \Delta y_0$

Similarly, $\Delta^3 y_{-1} = \Delta^2 y_0 - \Delta^2 y_{-1}$, therefore $\Delta^2 y_0 = \Delta^2 y_{-1} + \Delta^3 y_{-1}$

In general, $\Delta^n y_0 = \Delta^n y_{-1} + \Delta^{n+1} y_{-1}$ (1)

5.5.1 Gauss Forward Interpolation Formula

This method is used to interpolate the values of the function for the value of p such that $0<p<1$.

Newton's forward interpolation formula is

$$\phi(x) = y_p = y_0 + \frac{p}{1!}\Delta y_0 + \frac{p(p-1)}{2!}\Delta^2 y_0 + \frac{p(p-1)(p-2)}{3!}\Delta^3 y_0 \ldots\ldots + \frac{p(p-1)\ldots\ldots\ldots(p-n+1)}{n!}\Delta^n y_0$$

where $p = \dfrac{x - x_0}{h}$

Through equation (1),

For $n = 2$; $\Delta^2 y_0 = \Delta^2 y_{-1} + \Delta^3 y_{-1}$

For $n = 3$; $\Delta^3 y_0 = \Delta^3 y_{-1} + \Delta^4 y_{-1}$

For $n = 4$; $\Delta^4 y_0 = \Delta^4 y_{-1} + \Delta^5 y_{-1}$ .... and so on.

Put these above equations in Newton's forward interpolation formula as

$$y_p = y_0 + \frac{p}{1!}\Delta y_0 + \frac{p(p-1)}{2!}[\Delta^2 y_{-1} + \Delta^3 y_{-1}] + \frac{p(p-1)(p-2)}{3!}[\Delta^3 y_{-1} + \Delta^4 y_{-1}]$$

$$\ldots\ldots\ldots + \frac{p(p-1)\ldots\ldots(p-n+1)}{n!}[\Delta^n y_{-1} + \Delta^{n+1} y_{-1}]$$

$$y_p = y_0 + \frac{p}{1!}\Delta y_0 + \frac{p(p-1)}{2!}\Delta^2 y_{-1} + \frac{p(p-1)}{2!}\left[1 + \frac{p-2}{3}\right]\Delta^3 y_{-1} + \frac{p(p-1)(p-2)}{3!}\left[1 + \frac{p-3}{4}\right]\Delta^4 y_{-1}] + .. \qquad (2)$$

[Considering upto 5th term]

Also we have

$\Delta^5 y_{-2} = \Delta^4 y_{-1} - \Delta^4 y_{-2}$; or $\Delta^4 y_{-1} = \Delta^4 y_{-2} + \Delta^5 y_{-2}$

In general; $\Delta^n y_{-1} = \Delta^n y_{-2} + \Delta^{n+1} y_{-2}$ (3)

Put equation (3) in equation (2) as

$$y_p = y_0 + \frac{p}{1!}\Delta y_0 + \frac{p(p-1)}{2!}\Delta^2 y_{-1} + \frac{(p+1)p(p-1)}{3!}\Delta^3 y_{-1} + \frac{(p+1)p(p-1)(p-2)}{4!}[\Delta^4 y_2 + \Delta^5 y_{-2}] + \ldots$$

$$y_p = y_0 + p\,\Delta y_0 + \frac{p(p-1)}{2!}\Delta^2 y_{-1} + \frac{(p+1)p(p-1)}{3!}\Delta^3 y_{-1} + \frac{(p+1)p(p-1)(p-2)}{4!}\Delta^4 y_{-2} + \ldots$$

This above equation is called Gauss's forward interpolation formula. It employs odd differences just below the central line and even difference on the central line as shown below.

$$\begin{array}{ccccccccc} y_0 & \cdots\cdots & \Delta^2 y_{-1} & \cdots\cdots & \Delta^4 y_{-2} & \cdots\cdots & \Delta^6 y_{-3} & \cdots\cdots & \text{central line} \\ & \searrow \nearrow & & \searrow \nearrow & & \searrow \nearrow & & \searrow & \\ & \Delta y_0 & \cdots\cdots & \Delta^3 y_{-1} & \cdots\cdots & \Delta^5 y_{-2} & \cdots\cdots & \Delta^7 y_{-3} & \cdots \end{array}$$

5.5.2 Gauss Backward Interpolation Formula

This method is used when x ranges $\left[\frac{1}{2}, 0\right]$, and used to interpolate the values of y for a negative value of p lying between –1 and 0.

Newton's forward interpolation formula is

$$\phi(x) = y_p = y_0 + \frac{p}{1!}\Delta y_0 + \frac{p(p-1)}{2!}\Delta^2 y_0 + \frac{p(p-1)(p-2)}{3!}\Delta^3 y_0 \cdots\cdots + \frac{p(p-1)\cdots\cdots(p-n+1)}{n!}\Delta^n y_0$$

$$y_p = y_0 + \frac{p}{1!}\Delta y_0 + \frac{p(p-1)}{2!}\Delta^2 y_0 + \frac{p(p-1)(p-2)}{3!}\Delta^3 y_0 + \cdots\cdots \quad (1)$$

[Considering upto 4th term]

where $p = \frac{x - x_0}{h}$

Through equation (1),

For n = 2; $\Delta y_0 = \Delta^2 y_{-1} + \Delta^2 y_{-1}$

For n = 3; $\Delta^2 y_0 = \Delta^2 y_{-1} + \Delta^3 y_{-1}$

For n = 4; $\Delta^3 y_0 = \Delta^3 y_{-1} + \Delta^4 y_{-1}$ .... and so on.

General form: $\Delta^n y_0 = \Delta^n y_{-1} + \Delta^{n+1} y_{-1}$ (2)

Put these above equations in Newton's forward interpolation formula as

$$y_p = y_0 + p\,[\Delta y_{-1} + \Delta^2 y_{-1}] + \frac{p(p-1)}{2!}[\Delta^2 y_{-1} - \Delta^3 y_{-1}] + \frac{p(p-1)(p-2)}{3!}[\Delta^3 y_{-1} + \Delta^4 y_{-1}]$$

$$+ \frac{(p-1)p(p-2)(p-3)}{4!}[\Delta^3 y_{-1} + \Delta^4 y_{-1}] + \cdots\cdots$$

$$y_p = y_0 + p\,\Delta y_{-1} + \left[p + \frac{p(p-1)}{2}\right]\Delta^2 y_{-1} + \left[\frac{p(p-2)}{2} + \frac{p(p-1)(p-2)}{6}\right]\Delta^3 y_{-1} +$$

$$\left[\frac{p(p-1)(p-2)}{6} + \frac{p(p-1)(p-2)(p-3)}{24}\right]\Delta^4 y_{-1} + \ldots$$

$$y_p = y_0 + p\,\Delta y_{-1} +$$

$$\left[\frac{p(p+1)}{2!}\right]\Delta^2 y_{-1} + \left[\frac{(p-1)p(p+1)}{3!}\right]\Delta^3 y_{-1} + \left[\frac{(p-2)(p-1)p(p+1)}{4!}\right]\Delta^4 y_{-1} + \quad (3)$$

Also through equation (2), we have

For n = 3; $\Delta^3 y_{-1} = \Delta^3 y_{-2} + \Delta^4 y_{-2}$

For n = 4; $\Delta^4 y_{-1} = \Delta^4 y_{-2} + \Delta^5 y_{-2}$............. and so on.

General form: $\Delta^n y_{-1} = \Delta^n y_{-2} + \Delta^{n+1} y_{-2}$

Put these above equations in equation (3) as

$$y_p = y_0 + \frac{p}{1!}\Delta y_{-1} + \frac{p(p+1)}{2!}\Delta^2 y_{-1} + \frac{(p-1)p(p+1)}{3!}[\Delta^3 y_{-2} + \Delta^4 y_{-2}] +$$

$$\frac{(p-2)(p-1)p(p+1)}{4!}[\Delta^4 y_2 + \Delta^5 y_{-2}] + \ldots$$

$$y_p = y_0 + p\,\Delta y_0 + \frac{p(p+1)}{2!}\Delta^2 y_{-1} + \frac{(p-1)p(p+1)}{3!}\Delta^3 y_{-2} +$$

$$\frac{(p-2)(p-1)p(p+1)}{4!}\Delta^4 y_{-2} + \ldots$$

This above equation is called Gauss's backward interpolation formula. It employs odd differences just above the central line and even difference on the central line as shown below.

Δy_{-1} .... $\Delta^2 y_{-2}$ .... $\Delta^4 y_{-3}$ .... $\Delta^6 y_{-4}$ ......

↗ ↘ ↗ ↘ ↗ ↘ ↗

y_0 ....$\Delta^2 y_{-1}$ .... $\Delta^4 y_{-2}$ .... $\Delta^6 y_{-3}$ .... central line

5.5.3 Stirling's Formula

This method is used when p is in range $\left[-\frac{1}{2},\frac{1}{2}\right]$ and gives the accurate results as compare to the Gauss's interpolation formula.

Gauss's forward interpolation formula is

$$y_p = y_0 + p\,\Delta y_0 + \frac{p\,(p-1)}{2!}\Delta^2 y_{-1} + \frac{(p+1)\,p\,(p-1)}{3!}\Delta^3 y_{-1} + \frac{(p+1)\,p\,(p-1)\,(p-2)}{4!}\Delta^4 y_{-2} + \dots$$

Gauss's backward interpolation formula is

$$y_p = y_0 + p\,\Delta y_0 + \frac{p\,(p+1)}{2!}\Delta^2 y_{-1} + \frac{(p-1)\,p\,(p+1)}{3!}\Delta^3 y_{-2} + \frac{(p-2)\,(p-1)\,p\,(p+1)}{4!}\Delta^4 y_{-2} + \dots$$

On taking the mean of above two formula's of Gauss's interpolation as

$$y_p = y_0 + p\left[\frac{\Delta y_0 + \Delta y_{-1}}{2}\right] + \frac{p}{2!}\left[\frac{p-1+p+1}{2}\right]\Delta^2 y_{-1} +$$

$$\frac{(p+1)\,p\,(p-1)}{3!}\left[\frac{\Delta^3 y_{-1} + \Delta^3 y_{-2}}{2}\right] + \frac{(p+1)\,p\,(p-1)}{4!}\left[\frac{p+2+p-2}{2}\right] + \Delta^4 y_{-2}$$

$$+ \dots$$

$$y_p = y_0 + p\left[\frac{\Delta y_0 + \Delta y_{-1}}{2}\right] + \frac{p^2}{2!}\Delta^2 y_{-1} +$$

$$\frac{p\,(p^2-1)}{3!}\left[\frac{\Delta^3 y_{-1} + \Delta^3 y_{-2}}{2}\right] + \frac{p^2\,(p^2-1)}{4!}\Delta^4 y_{-2} + \dots$$

The above equation is called as Stirling's formula.

5.5.4 Bessel's Formula

This method is used when p is in range $\left[\frac{1}{4},\frac{3}{4}\right]$ and odd differences below the central line and mean of even differences on and below the line.

X	y	1st difference, Δy	2nd difference, $\Delta^2 y$	3rd difference, $\Delta^3 y$	4th difference, $\Delta^4 y$	5th difference, $\Delta^5 y$
x_0-2h	y_{-2}					
		Δy_{-2}				
x_0-h	y_{-1}		$\Delta^2 y_{-2}$			
		Δy_{-1}		$\Delta^3 y_{-2}$		
x_0	y_0		$\Delta^2 y_{-1}$		$\Delta^4 y_{-2}$	
		Δy_0		$\Delta^3 y_{-1}$		$\Delta^5 y_{-2}$
x_0+h	y_1		$\Delta^2 y_0$		$\Delta^4 y_{-1}$	
		Δy_1		$\Delta^3 y_0$		
x_0+2h	y_2		$\Delta^2 y_1$			
		Δy_2				
x_0+3h	y_3					

Gauss's forward interpolation formula is

$$y_p = y_0 + p\,\Delta y_0 + \frac{p(p-1)}{2!}\Delta^2 y_{-1} + \frac{(p+1)p(p-1)}{3!}\Delta^3 y_{-1} + \frac{(p+1)p(p-1)(p-2)}{4!}\Delta^4 y_{-2} + \ldots \qquad (1)$$

Gauss's backward interpolation formula is

$$y_p = y_0 + p\,\Delta y_0 + \frac{p(p+1)}{2!}\Delta^2 y_{-1} + \frac{(p-1)p(p+1)}{3!}\Delta^3 y_{-2} + \frac{(p-2)(p-1)p(p+1)}{4!}\Delta^4 y_{-2} + \ldots$$

Shifting the origin on point backward through replacement of p by (p–1) and addition of 1 to subscript y, the resulting equation becomes

$$y_p = y_1 + (p-1)\Delta y_0 + \frac{p(p-1)}{2!}\Delta^2 y_0 + \frac{p(p-1)(p-2)}{3!}\Delta^3 y_{-1} + \frac{(p+1)p(p-1)(p-2)}{4!}\Delta^4 y_{-1} + \ldots \qquad (2)$$

On taking mean of equation (1) and (2), the resulting equation is

$$y_p = y_1 + \frac{1}{2}(y_0 + y_1) + \left(p - \frac{1}{2}\right)\Delta y_0 + \frac{p(p-1)}{2!}\cdot\frac{1}{2}(\Delta^2 y_{-1} + \Delta^2 y_0) + \frac{\left(p-\frac{1}{2}\right)p(p-1)}{3!}\Delta^3 y_{-1} + \frac{(p+1)p(p-1)(p-2)}{4!}\cdot\frac{1}{2}(\Delta^4 y_{-1} + \Delta^4 y_0) + \ldots$$

The above equation is called Bessel's formula.

5.5.5 Laplace-Everett's Formula

This method is used most and it involves only even differences on and below the central line. This formula can be converted to Bessel's formula through suitable arrangements. On comparison with Bessel's formula, it should be noted that Laplace-Everett's formula truncated after 2nd differences and Bessel's formula truncated after 3rd differences is same.

Gauss's forward interpolation formula is

$$y_p = y_0 + p\,\Delta y_0 + \frac{p\,(p-1)}{2!}\,\Delta^2 y_{-1} + p\,\Delta y_0 + \frac{p\,(p-1)}{2!}\,\Delta^3 y_{-1} +$$

$$\frac{(p+1)\,p\,(p-1)\,(p-2)}{4!}\,\Delta^4 y_{-2} + \frac{(p+2)\,(p+1)\,p\,(p-1)\,(p-2)}{5!}\,\Delta^5 y_{-2} + \ldots \quad (1)$$

Equation (1) changes on elimination of odd differences i.e., $\Delta y_0 = y_1 - y_0$, $\Delta^3 y_{-1} = \Delta^2 y_0 - \Delta^2 y_{-1}$, $\Delta^5 y_{-2} = \Delta^4 y_{-1} - \Delta^4 y_{-2}$, and so on.

$$y_p = y_0 + p\,(y_0 - y_1) + \frac{p\,(p-1)}{2!}\,\Delta^2 y_{-1} + \frac{(p+1)\,p\,(p-1)}{3!}\,(\Delta^2 y_0 - \Delta^2 y_{-1}) +$$

$$\frac{(p+1)\,p\,(p-1)\,(p-2)}{4!}\,\Delta^4 y_{-2} + \frac{(p+2)\,(p+1)\,p\,(p-1)\,(p-2)}{5!}\,(\Delta^4 y_{-1} - \Delta^4 y_{-2}) + \ldots$$

$$y_p = (1 - p)y_0 + py_1 - \frac{p\,(p-1)\,(p-2)}{3!}\,\Delta^2 y_{-1} + \frac{(p+1)\,p\,(p-1)}{3!}\,\Delta^2 y_0 -$$

$$\frac{(p+1)\,p\,(p-1)\,(p-2)\,(p-3)}{5!}\,\Delta^4 y_{-2} + \frac{(p+2)\,(p+1)\,p\,(p-1)\,(p-2)}{5!}\,\Delta^4 y_{-1} -$$

On replacement of $1 - p$ as q in above equation yielding

$$y_p = q\,y_0 + \frac{q^2\,(q^2 - 1^2)}{3!}\,\Delta^2 y_{-1} + \frac{q^2\,(q^2 - 1^2)\,(q^2 - 2^2)}{5!}\,\Delta^4 y_{-2} + \ldots$$

$$\Delta^4 y_{-2} + \ldots \frac{p^2\,(p^2 - 1^2)}{3!}\,\Delta^2 y_0 + \frac{p^2\,(p^2 - 1^2)\,(p^2 - 2^2)}{5!}\,\Delta^4 y_{-1} + \ldots$$

The above equation is called Laplace-Everett's formula.

5.6 Cubic Spline Interpolation

Only one polynomial is used to explain the complete range of data, in general. But in this method, different continuous polynomials are used to describe the function in each interval of known points, so this approximation is called the **piecewise polynomial approximation**. There will be n piecewise polynomials for n+1 data sets. In literature, splines of different degree are available. **Cubic spline interpolation** method interpolates a function between a given set of data points through piecewise smooth polynomials. In this, the curve passes through the given set of data points and the slope & its curvature are continuous at each point. This method is very powerful and widely used w.r.t accuracy in the formation of polynomials of low degree and is less oscillatory in nature. So this method is used in numerical differentiation, integration, solution of boundary value problems, plotting of 2 and 3 – dimensional graph.

An expression for 2nd derivative can be obtained that describe the behaviour of data most accurately within each interval. The 2nd derivatives of the spline is zero at the end points. Since these end conditions occur naturally in a beam model (in strength of materials), so the resulting curve is called **natural cubic spline**. The pins (data points) are called **knots** of the spline in a beam model.

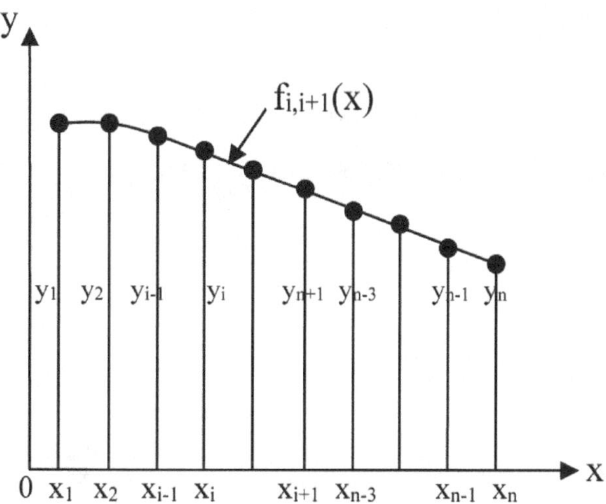

Fig. 5.1: Cubic spline interpolation

The above figure represents a cubic spline that spans n knots. Consider $f_{i,i+1}(x)$ is the cubic polynomial that spans the segment between knots i and i+1. It is noted that the spline is a piecewise cubic curve, assembled together form the n–1 cubics $f_{1,2}(x)$, $f_{2,3}(x)$, ..., $f_{n-1,n}(x)$, all of which have different coefficients. Denoting the 2nd derivative of the spline at knot i by k_i, the continuity of 2nd derivatives requires that; $f''_{i-1,i}(x_i) = f''_{i,i+1}(x_i) = k_i$ (1)

Where k_i is unknown, except for; $k_1 = k_n = 0$

It is known that the expression for $f''_{i,i+1}(x)$ is linear and the starting point for obtaining the coefficients of $f_{i,i+1}(x)$ is $f''_{i,i+1}(x)$. So, we can write using Lagrange's two-point interpolation,

$$f''_{i,i+1}(x) = k_i\ l_i(x) + k_{i+1}\ l_{i+1}(x) \quad (2)$$

Where $l_i(x) = \dfrac{x - x_{i+1}}{x_i - x_{i+1}}$ and $l_{i+1}(x) = \dfrac{x - x_i}{x_{i+1} - x_i}$

Hence $f''_{i,i+1}(x) = \dfrac{k_i(x - x_{i-1}) - k_{i+1}(x - x_i)}{x_i - x_{i+1}}$ (3)

On integrating equation (3) w.r.t x, yields

$$f''_{i,i+1}(x) = \frac{k_i(x - x_{i-1})^3 - k_{i+1}(x - x_i)^3}{6(x_i - x_{i+1})} + A(x - x_{i+1}) - B(x - x_i) \quad (4)$$

$$f''_{i,i+1}(x) = \frac{k_i(x - x_{i-1})^3 - k_{i+1}(x - x_i)^3}{6(x_i - x_{i+1})} + Cx + D \quad (5)$$

Where A and B are integration constants, C = A – B and D = $-Ax_{i+1} + Bx_i$

Through equation (4), apply the condition $f_{i,i+1}(x_i) = y_i$;

$$\frac{k_i(x - x_{i+1})^3}{6(x_i - x_{i-1})} + A(x - x_{i+1}) = y_i \quad (6)$$

Hence, $A = \dfrac{y_i}{x_i - x_{i+1}} - \dfrac{k_i(x_i - x_{i+1})}{6}$ (7)

Similarly, applying the condition $f_{i,i+1}(x_{i+!}) = y_{i+1}$; $B = \dfrac{y_{i+1}}{x_i - x_{i+1}} - \dfrac{k_{i+1}(x_i - x_{i+1})}{6}$

(8)

From equations (7) and (8),

$$f_{i,i+1}(x) = \frac{k_i}{6}\left[\frac{(x - x_{i+1})^3}{x_i - x_{i+1}} - (x - x_{i+1})(x_i - x_{i+1})\right] - \frac{k_{i+1}}{6}\left[\frac{(x - x_i)^3}{x_i - x_{i+1}} - (x - x_i)(x_i - x_{i+1})\right] + \frac{y_i(x - x_{i+1}) - y_{i+1}(x - x_i)}{x_i - x_{i+1}}$$

(9)

It is noted that the 2nd derivatives k_i of the spline at the interior knots are found from the slope continuity conditions; $f'_{i-1}(x_i) = f'_{i,i+1}(x_i)$, i = 1,2,3,......., $n-1$

(10)

Applying the conditions through equation (10) in equation (9) and after some mathematical simplification, the resulting simultaneous equations (11) below are:

$$k_{i-1}(x_{i-1}-x_i)+2k_i(x_{i-1}-x_{i+1})+k_{i+1}(x_i-x_{i+1})=6\left[\frac{y_{i-1}-y_i}{x_{i-1}-x_i}-\frac{y_i-y_{i+1}}{x_i-x_{i+1}}\right];i=1,2,3,....,n-1$$

If the data points are spaced equally at h intervals, then $h = x_{i-1} - x_i = x_i - x_{i+1}$
And equation (11) becomes

$$k_{i-1}+4k_i+k_{i+1}=\frac{6}{h^2}\left[y_{i-1}-2y_i+y_{i+1}\right];i=1,2,3,....,n-1 \qquad (12)$$

There are 2 boundary conditions normally used and they are

(1) Natural (or free) boundary condition: The 2nd derivatives of the data at the end points x_0 and x_n are assumed to be zero. The condition is known as the natural boundary condition. The polynomials resulting from this condition are called natural cubic splines. They may not provide very accurate values close to the boundaries, but they are accurate enough in the interior region.

(2) Clamped boundary condition: When the 1st derivative of the data are known at the end points x_0 and x_n, the corresponding boundary condition are known. This condition is called as clamed boundary condition.

Example 5.11: Find the natural cubic spline interpolation at x = 3.4. The data points are

x:	1	2	3	4	5
y:	13	15	12	9	13

Solution: For equally spaced knots, the equation for the curvatures is equation (12), in which $k_1 = k_5$ and h = 1.

Hence,

$4k_2 + k_3 = 6\ [13 - 2\ (15) + 12] = -30$

$k_2 + 4k_3 + k_4 = 6\ [15 - 2\ (12) + 9] = 0$

$k_3 + 4k_4 = 6\ [12 - 2\ (9) + 13] = 42$

On solving above equations; $k_2 = -7.286$, $k_3 = -0.857$, $k_4 = 10.714$

The interpolant between knots 2 and 3 is given through equation (9) as

$$f_{i,i+1}(x) = \frac{k_i}{6}\left[\frac{(x-x_{i+1})^3}{x_i - x_{i+1}} - (x-x_{i+1})(x_i - x_{i+1})\right] - \frac{k_{i+1}}{6}\left[\frac{(x-x_i)^3}{x_i - x_{i+1}} - (x-x_i)(x_i - x_{i+1})\right]$$
$$+ \frac{y_i(x-x_{i+1}) - y_{i+1}(x-x_i)}{x_i - x_{i+1}}$$

$$f_{3,4}(x) = \frac{k_3}{6}\left[\frac{(x-x_4)^3}{x_3 - x_4} - (x-x_4)(x_3 - x_4)\right] - \frac{k_4}{6}\left[\frac{(x-x_3)^3}{x_3 - x_4} - (x-x_3)(x_3 - x_4)\right]$$
$$+ \frac{y_3(x-x_4) - y_4(x-x_3)}{x_3 - x_4}$$

$$f_{3,4}(x) = \frac{-0.857}{6}\left[\frac{(3.4-4)^3}{3-4} - (3.4-4)(3-4)\right] - \frac{10.714}{6}\left[\frac{(3.4-3)^3}{3-4} - (3.4-4)(3-4)\right]$$
$$+ \frac{12(3.4-4) - 9(3.4-3)}{3-4} = 10.2552$$

Exercise

5.1. Explain interpolation.

5.2. Differentiate interpolation and extrapolation, with the help of example.

5.3. What are finite difference operators? Explain each one in detail.

5.4. Construct the forward difference table and the horizontal (or backward) table for the following data

(a)

x	1	2	3	4	5
y = f(x)	4	6	9	12	17

(b)

X	0	10	20	30
y = f(x)	0	0.174	0.347	0.518

5.5. Construct a central difference table for $y = f(x) = x^3 + 2x + 1$ for x = 1, 2, 3, 4, 5

5.6. Obtain the backward differences for the function $f(x) = x^3$ from x =1 to 1.05 to two decimals chopped.

5.7. (a) Given $x_0 = 1$, $x_1 = 5$, $x_2 = 10$, $x_3 = 30$, and $x_4 = 40$, then find $\Delta^4 x_0$.

(b) Given $x_0 = 5$, $x_1 = 24$, $x_2 = 81$, $x_3 = 200$, $x_4 = 100$, and $x_5 = 8$, then find $\Delta^5 x_0$.

5.8. Estimate the missing term in the following table

(a)

x	1	2	3	4	5
y = f(x)	5	14	?	74	137

(b)

x	1	2	3	4	5
y = f(x)	8	17	38	?	140

(c)

x	0	1	2	3	4
y = f(x)	3	2	3	?	11

(d)

X	0	1	2	3	4
y = f(x)	–31	–35	?	5	133

5.9. Express the following questions in the factorial notation. Take interval of differencing as equal to 1.

(a) $y = f(x) = 3x^3 + x^2 + x + 1$

(b) $y = f(x) = x^4 - 5x^3 + 3x + 4$

5.10. The following table of a polynomial of 3rd degree. It is given that f(3) is in error. Correct the error.

x	0	1	2	3	4	5	6
y	1	2	33	254	1054	3126	7777

5.11. Give relationship between the operators.

5.12. Given that $(12600)^{0.5} = 112.24972$, $(12610)^{0.5} = 112.19426$, $(12620)^{0.5} = 112.33877$, $(12630)^{0.5} = 112.38327$. Find the value of $(12616)^{0.5}$.

5.13. Evaluate $y = e^{2x}$ for x = 0.25 from the data in the following table

x	0.2	0.3	0.4	0.5	0.6
e^{2x}	1.49182	1.82212	2.22554	2.71828	3.32012

5.14. A 2nd degree polynomial passes through the points (1, –1), (2, –2), (3, –1) and (4, 2). Find the polynomial.

5.15. The values of sin x are given below for different values of x, find the value of sin 45°.

x	40	45	50	55	60
y = sin x	0.6428	0.7071	0.7660	0.8192	0.8660

5.16. Calculate the total profits between the years 1990 – 2002. The profits of a company (in thousands) are given below:

Year (x)	1990	1993	1996	1999	2002
Profit, y = f(x)	120	100	111	108	99

5.17. Find f(0.5) through Newton's forward difference formula. If f(x) is known at the following data points

x_i	0	1	2	3	4
f_i	1	7	23	55	109

5.18. Find f(0.15) through Newton's backward difference formula from the following data points

x	f(x)	∇f	$\nabla^2 f$	$\nabla^3 f$	$\nabla^4 f$
0.1	0.09983				
		0.09884			
0.2	0.19867		-0.00199		
		0.09685		-0.00156	
0.3	0.29552		-0.00355		0.00121
		0.0939		-0.00035	
0.4	0.38942		-0.0039		
		0.09			
0.5	0.97943				

5.19. Use Gauss forward formula to find x = 30 given the following table of values

x	21	25	29	33	37
y	18.4708	17.8144	17.1070	16.3432	15.5154

5.20. From the following table, estimate the number of students who obtained marks in English between 75 and 80.

Marks	35–45	45–55	55–65	65–75	75–85
Number of students	20	40	60	60	20

5.21. Use Newton's forward interpolation formula to find the value of cos 52^0 through the following table

X	45^0	50^0	55^0	60^0
y = f(x)	0.7071	0.6428	0.5736	0.5

5.22. Using Lagrange's interpolation formula,

(a) Find a polynomial which passes though the points (0, –12), (1, 0), (3, 6) and (4, 12).

(b) Find the value of y corresponding to x = 10 through the table

x	5	6	9	11
y = f(x)	380	–2	196	308

5.23. Use Gauss backward interpolation formula, to find the sales for the year 1986 from the following table

Year	1951	1961	1971	1981	1991	2001
Sales (in thousands)	13	17	22	28	41	53

5.24. Apply Bessel's interpolation formula to obtain y_{25} and it is given that $y_{20} = 2860$, $y_{24} = 3167$, $y_{28} = 3555$, $y_{32} = 4112$.

5.25. Apply Stirling's interpolation formula to obtain y_{28} and it is given that $y_{20} = 48324$, $y_{25} = 47354$, $y_{30} = 46267$, $y_{35} = 44978$, and $y_{40} = 43389$.

5.26. Apply Laplace-Everett's interpolation formula to obtain y at x = 1.60, through the following table

X	1	1.25	1.50	1.75	2	2.25
y = f(x)	1.0543	1.1281	1.2247	1.3219	1.4243	1.4987

5.27. Find the form of a function f(x) under the suitable assumption from the following data

x	0	1	2	5
y = f(x)	2	3	12	147

5.28. Fit the cubic spline curve that passes through the points as mentioned below

x	0	1	2	3
y = f(x)	0	0.5	2	1.5

5.29. Fit the data in the table with cubic spline and find the value at x = 5.

i	1	2	3	4
x	3	4.5	7	9
y = f(x)	2.5	1	2.5	0.5

5.30. Develop a natural cubic spline for the following data

X	3	4	5	6	7
y = f(x)	3.7	3.9	3.9	4.2	5.7

Find f′(3.4), f′(5.2) and f′(5.6).

5.31. Determine the cubic spline interpolation at x = 2.6 based on the data:

x	0	1	2	3
y = f(x)	1	1	0.5	0

Given the end conditions as $f'_{1,2}(0) = 0$ (zero slope).

6

Curve Fitting

6.1 Introduction

Scientists and engineers have an experimental data of various variables in an experiment. To know the facts and to formulate related theories, they need exact relationship between such variables. So, they require the exact function which defines the outputs with introducing the inputs. For example,

Ohm's law: which states that voltage of a circuit is proportional to the current flowing in it i.e.

$V \propto I \rightarrow V = IR;$

V is the function which gives the various output values of voltage with the input values of the current.

Kepler's Law: In 1601, the German astronomer: Johannes Kepler formulated the third law of planetary motion i.e.

$T = C\,x^{\frac{3}{2}};$

where x is the distance to the sun measured.

Thus, in applications of science, there are mathematical relationships between the experimental quantities and it requires the establishment of their relationships in the form of mathematical equations. The process of establishing such relationships is known as curve fitting or regression analysis.

Let, the values of dependent variable y for the different values of independent variable x are given and it is required to establish their mathematical relationship as $y = f(x)$. Hence, the evaluation of curve for a given data is process of curve fitting. Also their mathematical relationships may be algebraic (linear or non-linear) or transcendental type (figure 6.1).

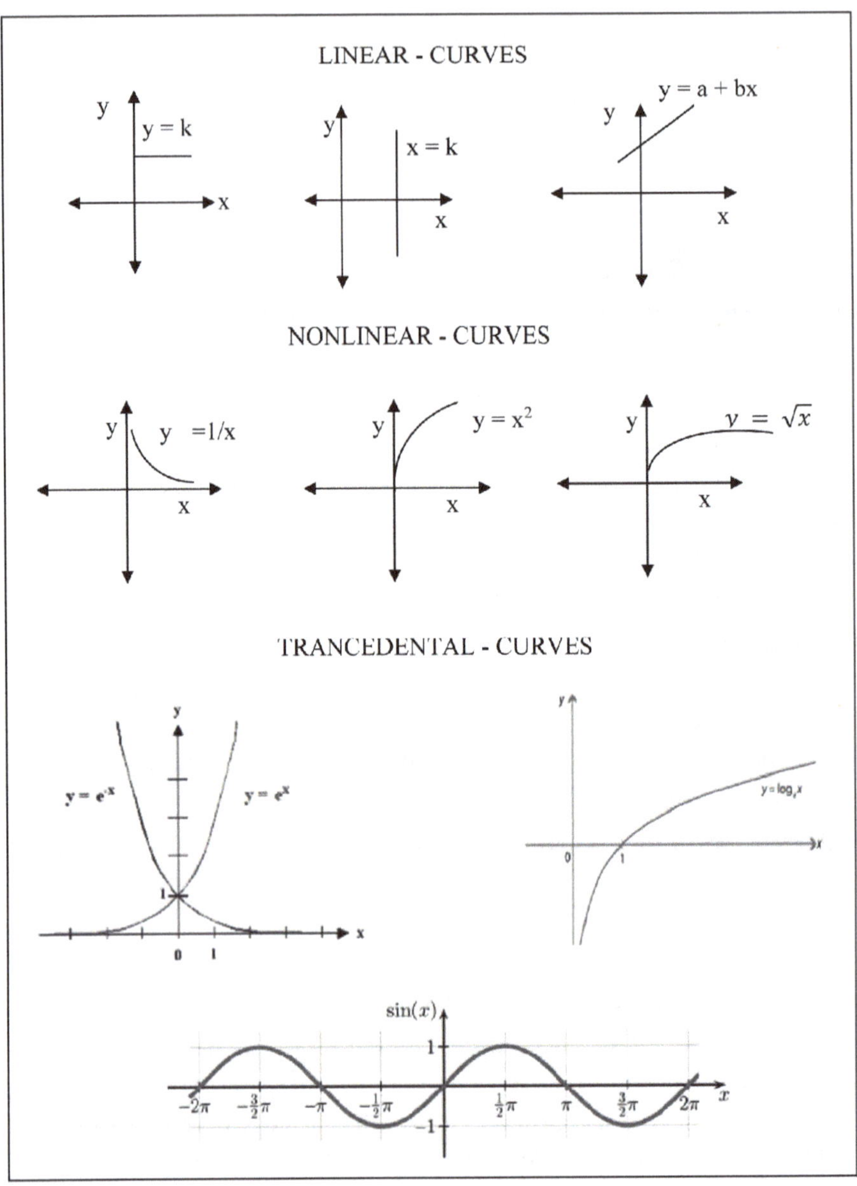

Fig. 6.1: different types of curves

6.2 Linear Equation

The curve $y = f(x) = a + bx$, is known as linear equation in two variables and it is a straight line when graphed in XY-Plane.

It has the following types:-

- **Horizontal line**

The graph of the equation; $y = \pm k$, is always a horizontal line:

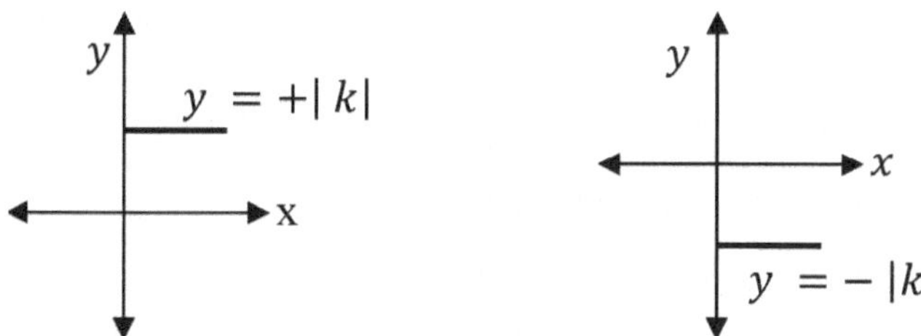

Fig. 6.2: Horizontal line

- **Vertical Line**

The graph of the equation; $x = \pm k$, is always a vertical line:

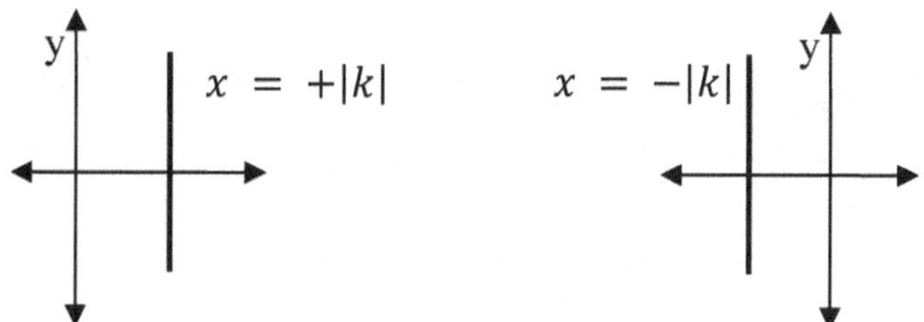

Fig. 6.3: Vertical line

- **Oblique line**

The graph of the equation; $y = a + bx$, is slanting line:

where a is the intercept on y-axis and b is the slope of this line.

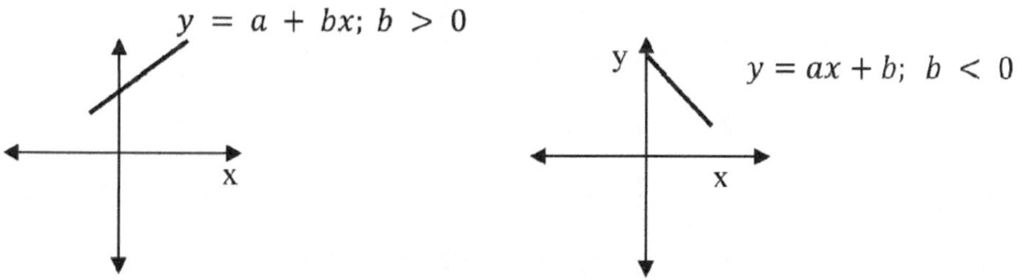

Fig. 6.4: Oblique line

Example 6.1: Draw the line $y = 5$.

Solution:

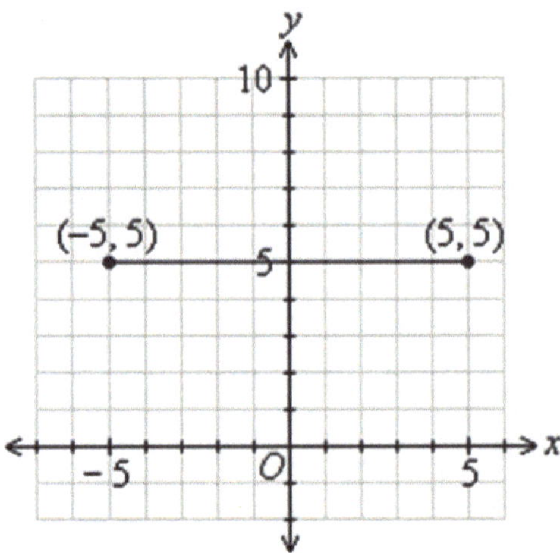

Example 6.2: Draw the line $x = 3$.

Solution:

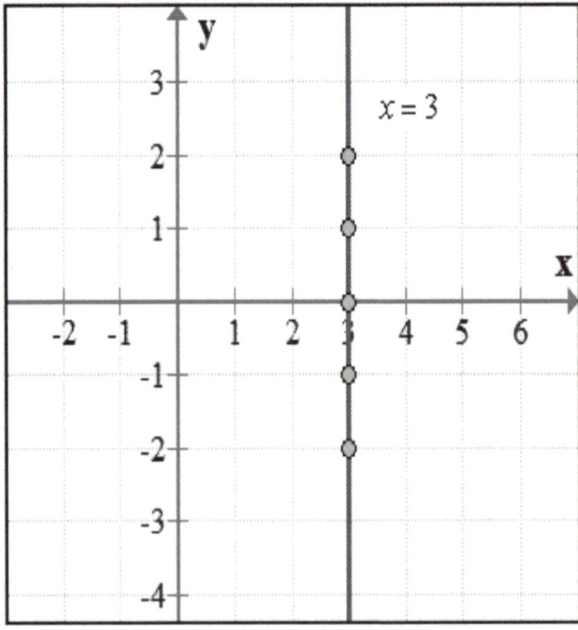

Example 6.3: Draw the line x + 2y = 7.

Solution: put $x = 0$;

$0 + 2y = 7$

$\Rightarrow y = \frac{7}{2}$

$\Rightarrow y = 3.5$

and now put $y = 0$;

$x + 0 = 7$

$\Rightarrow x = 7$

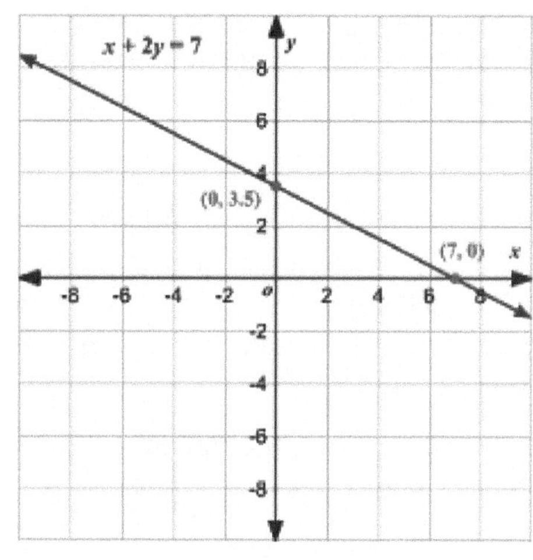

6.3 Curve Fitting with Linear Equation

Consider that an experiment produces a set of data points (x_1,y_1), (x_2, y_2), ..., (x_n, y_n) as shown in figure such that $\{x_i\ (i = 1,2 \ldots, n)\}$ are the distinct values. Now, there is a need to formulate a curve y = f(x) that relates these variables. In this section, we are using a class of linear functions with functional form as Y = a + bx which is shown in figure 6.5 with solid line.

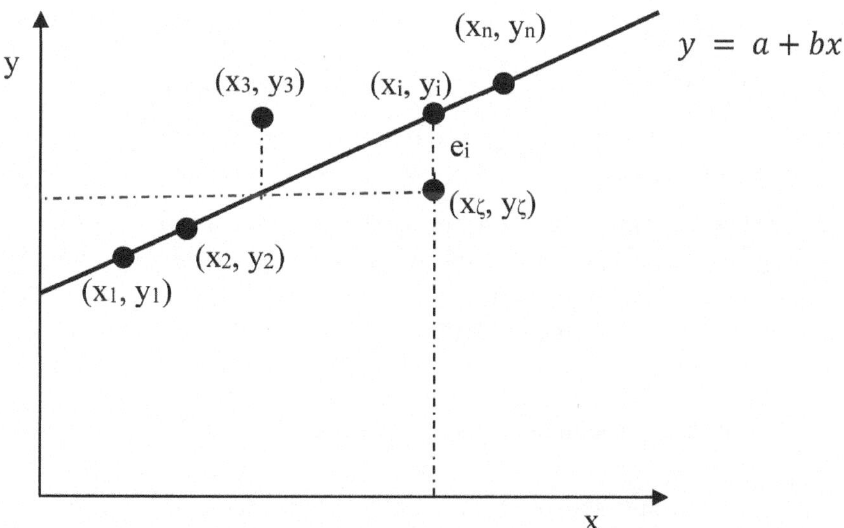

Fig. 6.5: Fitted straight line for given data points

The plotted points are the actual data of an experiment and the solid line is the fitted line for these actual data points. The points which lie on the fitted line are the accurate values while the points which do not lie have errors. The vertical distances of such points from the fitted line are considered as their error values or residuals and these residual are denoted by e_i i.e.

$e_i = y_i - y_i = y_i - (a + bx_i)$

The fitted line is the 'best' fit for the experimental data if their error values are negligible. Thus, it is always required to use a methodology which reduces the error values as small as possible. There are various approaches that could be tried for fulfillment the required results. Some are explained in the following way:

(a) To minimize the sum of errors i.e.

$$\text{Min}\sum e_i = \sum (y_i - (a + bx_i))$$

(b) To minimize the sum of absolute values of errors i.e.

$$\text{Min}\sum |e_i| = \sum |y_i - (a + bx_i)|$$

(c) To minimize the sum of square of errors i.e.

$$\text{Min}\sum e_i^2 = \sum (y_i - (a + bx_i))^2$$

6.4 Criteria for a Best Fit

A fitted curve is called as best fit for the given set of data points if it has the following properties:

- A fitted curve pass through the maximum number of given data points.
- If a fitted curve does not pass through some given data points then the errors or residuals to corresponding points should be least as possible.

Thus fitted curve is the 'best' fit for the experimental data if their error values are negligible. Thus, it is always required to use a methodology which reduces the error values as small as possible. There are various approaches that could be tried for fulfillment the required results. Some are explained in the following way:

(a) To minimize the sum of errors i.e. $\text{Min}\sum e_i$

(b) To minimize the sum of absolute values of errors i.e. $\text{Min}\sum |e_i|$

(c) To minimize the sum of square of errors i.e. $\text{Min}\sum e_i^2$

6.5 Linear Least-Square Regression

As described in section 6.3, there are the various approaches to fit the linear function for an experimental data. The common three approaches are also explained. From these approaches, first two approaches did not provide a unique fit line but the third approach yield a unique line. This methodology used for best fit is known as least squares method.

Let us denote the sum of square of error values by E i.e.

$$E = \sum e_i^2 = \sum \left(y_i - \left(a + bx_i\right)\right)^2 \quad (6.5.1)$$

Since fitted line is evaluated by using the functional relationship as $Y = a + bx$; thus we have to evaluate the coefficients a and b. These coefficients are to be chosen in such a way that the value of E should be minimum. Thus the necessary conditions for minimizing E by using the results of calculus, can be defined as:

$$\frac{\partial E}{\partial a} = 0 \text{ and } \frac{\partial E}{\partial b} = 0$$

Now, differentiate the equation (6.5.1) partially w.r.t. a; we get:

$$\frac{\partial E}{\partial a} = 2\sum \left(y_i - \left(a + bx_i\right)\right)(-1)$$

Since $\frac{\partial E}{\partial a} = 0$, therefore we have

$$2\sum \left(y_i - \left(a + bx_i\right)\right) = 0$$

$$\Rightarrow \sum \left(y_i - \left(a + bx_i\right)\right) = 0$$

$$\Rightarrow \sum \left(y_i\right) - \sum a - \sum bx_i = 0$$

$$\Rightarrow \sum \left(y_i\right) = \sum a + \sum bx_i$$

$$\Rightarrow \sum \left(y_i\right) = a\sum 1 + \sum bx_i$$

$$\Rightarrow \sum \left(y_i\right) = an + b\sum x_i \quad (6.5.2)$$

Now, similarly differentiate the equation (6.5.1) partially w.r.t. b; we get:

$$\frac{\partial E}{\partial b} = 2\sum \left(y_i - \left(a + bx_i\right)\right)\left(-x_i\right)$$

Since $\frac{\partial E}{\partial b} = 0$, therefore we have

$$2\sum(y_i-(a+bx_i))(-x_i)=0$$

$$\Rightarrow\sum(y_i-(a+bx_i))(-x_i)=0$$

$$\Rightarrow\sum(x_iy_i)-\sum ax_i-\sum bx_i^2=0$$

$$\Rightarrow\sum(x_iy_i)=\sum ax_i+\sum bx_i^2$$

$$\Rightarrow\sum(x_iy_i)=a\sum x_i+b\sum x_i^2 \qquad (6.5.3)$$

After solving equations (6.5.2) and (6.5.3), we get the values of a and b as the following:

$$a=\frac{\left(\sum x_i^2\right)\left(\sum y_i\right)-\left(\sum x_i\right)\sum(x_iy_i)}{n\sum x_i^2-\left(\sum x_i\right)^2},\ b=\frac{n\sum(x_iy_i)-\left(\sum x_i\right)\left(\sum y_i\right)}{n\sum x_i^2-\left(\sum x_i\right)^2} \qquad (6.5.4)$$

Now; from an experimental data, we can find the values of $\left(\sum x_i\right),\left(\sum y_i\right),\sum(x_iy_i),\left(\sum x_i^2\right)$ and after substituting these values in the equations (6.5.4), the values of a and b can be evaluated.

Hence, the linear fit can be formulated from the given experimental data points $(x_1,y_1),(x_2,y_2),\ldots,(x_n,y_n)$.

6.6 Method of Least Squares Fitting: Straight Line, 2nd Degree Parabola, Exponential Curves

Method of least squares is widely used for fitting the algebraic and transcendental curves to the data points. In this section, we are explaining the least squares methodology for fitting straight line, second degree parabola and exponential curves for the given data points in the following way:

6.6.1 Fitting of straight line

Method of fitting straight line for given data points is already explained in section 6.5.

Example 6.4: Fit a least square straight line to the following set of data:

x:	1	2	3	4	5
y:	3	4	6	7	8

Solution: Let us assume the straight line and use the equations (6.5.4) to evaluate the values a and b.

x_i	y_i	x_i^2	y_i^2	x_iy_i
1	3	1	9	3
2	4	4	16	8
3	6	9	36	18
4	7	16	49	28
5	8	25	64	40
$\Sigma x_i = 15$	$\Sigma y_i = 28$	$\Sigma x_i^2 = 55$	$\Sigma y_i^2 = 174$	$\Sigma x_iy_i = 97$

n = 5

$$a = \frac{n\sum(x_i^2)\left(\sum y_i\right) - \left(\sum x_i\right)\left(\sum x_i y_i\right)}{n\sum x_i^2 - \left(\sum x_i\right)^2} = \frac{55\times 28 - 15\times 97}{5\times 55 - 15\times 15} = 1.7$$

$$b = \frac{n\sum(x_i y_i) - \left(\sum x_i\right)\left(\sum y_i\right)}{n\sum x_i^2 - \left(\sum x_i\right)^2} = \frac{5\times 97 - 15\times 28}{5\times 55 - 15\times 15} = 1.3$$

Hence, $y = 1.7 + 1.3x$ is fitted line for the given data.

6.6.2 Fitting of second degree parabola

Consider a set of data points as $(x_1, y_1), (x_2, y_2), \ldots, (x_n, y_n)$ and we want to fit the second degree parabolic curve as:

$$Y = a + bx + cx^2$$

The residuals e_i are

$$e_i = y_i - y_i = y_i - (a + bx_i + cx^2)$$

and the total sum of squares of residuals is E which can be defined as:

$$E = \Sigma e_i^2 = \Sigma\left(y_i - (a + bx_i + cx^2)\right)^2 \tag{6.6.2}$$

According to method of least squares; the coefficients a, b, c are to be chosen in such a way that the value of E is minimized. Thus the necessary conditions, for minimizing E by using the results of calculus, can be defined as:

$$\frac{\partial E}{\partial a} = 0,\quad \frac{\partial E}{\partial b} = 0 \quad \text{and} \quad \frac{\partial E}{\partial c} = 0$$

Now, differentiate the equation (6.6.2.1) partially w.r.t. a; we get:

$$\frac{\partial E}{\partial a} = 2\sum\left(y_i - \left(a + bx_i + cx_i^2\right)\right)(-1)$$

Since $\frac{\partial E}{\partial a} = 0$, therefore we have

$$2\sum\left(y_i-\left(a+bx_i+cx_i^2\right)\right)=0$$

$$\Rightarrow\sum\left(y_i-\left(a+bx_i+cx_i^2\right)\right)=0$$

$$\Rightarrow\sum(y_i)-\sum a-\sum bx_i-\sum cx_i^2=0$$

$$\Rightarrow\sum(y_i)=\sum a+\sum bx_i+\sum cx_i^2$$

$$\Rightarrow\sum(y_i)=a\sum 1+\sum bx_i+\sum cx_i^2$$

$$\Rightarrow\sum(y_i)=an+b\sum x_i+c\sum x_i^2 \qquad (6.6.2.2)$$

Now, similarly differentiate the equation (6.6.2.1) partially w.r.t. b; we get:

$$\frac{\partial E}{\partial b}=2\sum\left(y_i-\left(a+bx_i+cx_i^2\right)\right)(-x_i)$$

Since $\frac{\partial E}{\partial b}=0$, therefore we have

$$2\sum\left(y_i-\left(a+bx_i+cx_i^2\right)\right)(-x_i)=0$$

$$\Rightarrow\sum\left(y_i-\left(a+bx_i+cx_i^2\right)\right)(-x_i)=0$$

$$\Rightarrow\sum(x_iy_i)-\sum ax_i-\sum bx_i^2-\sum cx_i^3=0$$

$$\Rightarrow\sum(x_iy_i)=\sum ax_i+\sum bx_i^2+\sum cx_i^3$$

$$\Rightarrow\sum(x_iy_i)=a\sum x_i+b\sum x_i^2+c\sum x_i^3 \qquad (6.6.2.3)$$

Now, similarly differentiate the equation (6.6.2.1) partially w.r.t. c; we get:

$$\frac{\partial E}{\partial c}=2\sum\left(y_i-\left(a+bx_i+cx_i^2\right)\right)(-x_i^2)$$

Since $\frac{\partial E}{\partial b}=0$, therefore we have

$$2\sum\left(y_i-\left(a+bx_i+cx_i^2\right)\right)(-x_i^2)=0$$

$$\Rightarrow\sum\left(y_i-\left(a+bx_i+cx_i^2\right)\right)(-x_i^2)=0$$

$$\Rightarrow\sum(x_i^2y_i)-\sum ax_i^2-\sum bx_i^3-\sum cx_i^4=0$$

$$\Rightarrow \sum\left(x_i^2 y_i\right) = \sum a x_i^2 + \sum b x_i^3 + \sum c x_i^4$$

$$\Rightarrow \sum\left(x_i^2 y_i\right) = a\sum x_i^2 + b\sum x_i^3 + c\sum x_i^4 \qquad (6.6.2.4)$$

After solving equations (6.6.2.2), (6.6.2.3) and (6.6.2.4), we get the values of a, b and c.

Hence, the parabolic fit Y = a + bx + cx^2 can be formulated from the given experimental data points (x_1, y_1), (x_z, y_z),...., (x_n , y_n).

Example 6.5: Fit a least square parabola to the following set of data points:

x:	2	4	6	8	10
y:	4	10	40	65	110

Solution: Let us assume the parabolic curve Y = a + bx + cx^2 and use the equations (6.6.2.2), (6.6.2.3) and (6.6.2.4), to evaluate the values of *a,b* and c.

x_i	y_i	x_i^2	x_i^3	x_i^4	$x_i y_i$	$x_i^2 y_i$
2	4	4	8	16	8	16
4	10	16	64	256	40	160
6	40	36	216	1296	240	1440
8	65	64	512	4096	520	4160
10	110	100	1000	10000	1100	11000
Σx_i 30	Σy_i 299	$2\Sigma x_i^2$ 20	Σx_i^3 1800	Σx_i^4 15664	$\Sigma x_i y_i$ 1908	$\Sigma x_i^2 y_i$ 16776

N = 5

we got the following equations:

$299 = 5a + 30b + 220c;$

$1908 = 30a + 220b + 1800c;$

$16776 = 220a + 1800b + 15664c;$

on solving these equations, we got the values of *a, b* and *c* as:

$a = 324.2, b = -117.8, c = 10.05$

Hence $y = 324.2 - 117.8x + 10.05x^2$ is fitted parabola for the given data.

6.6.3 Fitting of exponential curve

Consider a set of data points as $(x_1, y_1), (x_2, y_2), \ldots, (x_n, y_n)$ and we want to fit an exponential curve as:

$$Y = a\, e^{Bx}$$

Firstly, we are taking logarithm both sides for this exponential curve; we get:

$$Y = \log a + Bx$$

Now, for simplicity let log a = A, Thus, we have

$$Y = A + Bx$$

For this curve, the residuals e_i are

$$e_i = y_i - Y_i = y_i - (A + Bx_i)$$

and the total sum of squares of residuals is E which can be defined as:

$$E = \sum e_i^2 = \sum \left(y_i - (A + Bx_i)\right)^2 \qquad (6.6.3.1)$$

According to method of least squares; the coefficients are to be chosen in such a way that the value of E is minimized. Thus the necessary conditions, for minimizing E by using the results of calculus, can be defined as:

$$\frac{\partial E}{\partial A} = 0 \text{ and } \frac{\partial E}{\partial B} = 0$$

Now, differentiate the equation (6.6.3.1) partially w.r.t. A; we get:

$$\frac{\partial E}{\partial A} = 2\sum \left(y_i - (A + Bx_i)\right)(-1)$$

Since $\frac{\partial E}{\partial A} = 0$, therefore we have

$$2\sum \left(y_i - (A + Bx_i)\right) = 0$$

$$\Rightarrow \sum \left(y_i - (A + Bx_i)\right) = 0$$

$$\Rightarrow \sum (y_i) - \sum A - \sum Bx_i = 0$$

$$\Rightarrow \sum (y_i) = \sum A + \sum Bx_i$$

$$\Rightarrow \sum (y_i) = A\sum 1 + \sum Bx_i$$

$$\Rightarrow \sum (y_i) = An + B\sum x_i \qquad (6.6.3.2)$$

Now, similarly differentiate the equation (6.6.3.1) partially w.r.t.B; we get:

$$\frac{\partial E}{\partial B} = 2\sum \left(y_i - (A + Bx_i)\right)(-x_i)$$

Since $\frac{\partial E}{\partial B} = 0$, therefore we have

$$2\sum\left(y_i-\left(A+Bx_i\right)\right)\left(-x_i\right)=0$$
$$\Rightarrow\sum\left(y_i-\left(A+Bx_i\right)\right)\left(-x_i\right)=0$$
$$\Rightarrow\sum\left(x_iy_i\right)-\sum Ax_i-\sum Bx_i^2=0$$
$$\Rightarrow\sum\left(x_iy_i\right)=\sum Ax_i+\sum Bx_i^2$$

$$\Rightarrow\sum\left(x_iy_i\right)=A\sum x_i+B\sum x_i^2 \qquad (6.6.3.3)$$

After solving equations (6.6.3.2) and (6.6.3.3), we get the values of a and b as the following:

$$A=\log a=\frac{\left(\sum x_i^2\right)\left(\sum y_i\right)-\left(\sum x_i\right)\sum\left(x_iy_i\right)}{n\sum x_i^2-\left(\sum x_i\right)^2};B=\frac{n\sum\left(x_iy_i\right)-\left(\sum x_i\right)\left(\sum y_i\right)}{n\sum x_i^2-\left(\sum x_i\right)^2}$$

(6.6.3.4)

In this way; we get the values of a, B.

Hence, the exponential fit Y = a e^{Bx} can be formulated from the given experimental data points (x_1, y_1), (x_2, y_2),…,(x_n, y_n).

Example 6.6: Fit a least square curve Y = a e^{Bx} to the following set of data:

x:	1	3	5	7	9
y:	3	8	16	20	25

Solution: To find the fitted curve Y=a e^{Bx}, we use the equations (6.6.3.4) to evaluate the values of *a* and b.

x_i	y_i	x_i^2	y_i^2	x_iy_i
1	3	1	3	3
3	8	9	24	8
5	16	25	80	18
7	20	49	140	28
9	25	81	225	40
$\Sigma x_i = 25$	$\Sigma y_i = 72$	$\Sigma x_i^2 = 165$	$\Sigma y_i^2 = 1354$	$\Sigma x_i y_i = 472$

N = 5

$$\log a=\frac{\left(\sum x_i^2\right)\left(\sum y_i\right)-\left(\sum x_i\right)\sum\left(x_iy_i\right)}{n\sum x_i^2-\left(\sum x_i\right)^2}=\frac{165\times72-25\times472}{5\times165-25\times25}=0.4;\quad a=e^{0.4}=1.5$$

$$B=\frac{n\sum\left(x_iy_i\right)-\left(\sum x_i\right)\left(\sum y_i\right)}{n\sum x_i^2-\left(\sum x_i\right)^2}=\frac{5\times472-25\times72}{5\times165-25\times25}=2.8$$

Hence, $y = 1.5\,e^{2.8x}$ is fitted parabola for the given values of data.

Exercise

6.1. Can a polynomial of any order necessarily fit any set of data?

6.2. Draw the lines: (a) Y = -5 (b) x = 4 (c) x + y =5

6.3. Find the least square fit line for the following experimental data:

x:	1	3	4	6	8	9	11
y:	1	2	5	4	4	8	7

6.4. Find the errors of least square fit line for the experimental data given in question 2.

6.5. In a certain city, efforts were introduced to reduce the polluted area and related results for ten months, are shown below:

Month:	1	2	3	4	5	6	7	8	9	10
Polluted Area (%):	10	9.5	8	8	8.5	7	6	6.5	5	4.5

Fit a least square second degree parabolic curve to this data and find out the approximate percentage of polluted area in eleventh month.

6.6. Fit an exponential curve $u = a\,e^{\beta t}$ to the following data:

u:	0.2	0.4	0.6	0.8	1.0	1.2	1.4
t:	10	18	26	30	36	50	65

6.7. Derive least square fit for the curves:

(a) $y = kx^2$ (b) $y = a + b \log_{10} x$

6.8. Find the least square linear and quadratic fit for the following data:

v	1	2	3	4	5
$h\ (v)$	2	4	3.5	6.75	8

6.9. An engineer wants to approximate the following experimental data:

u:	0	0.5	1	2.0
f(u):	2.95	3.9445	5.5043	11.7889

by using the function $f(u) = x\,e^u + y$. Determine the values of x and y by using least approximation.

6.10. Find the residuals of linear fit for the following data by using least square approximation:

x:	-2	-1	0	1	2
f(x):	6	3	2	2	1

7

Numerical Integration

7.1 Introduction

A Scottish mathematician David Gibb was firstly introduced a term 'Numerical Integration' in 1915. Like numerical differentiation, numerical integration is also a very useful tool for scientists and engineers especially when they have an experimental data for different variables without knowing any mathematical relationship between them. Beside it, there are also many other situations where it plays an important role for integrating the functions like:

- The integrand $g(x)$ of $I(x)=\int_a^b g(x)\,dx$ possess this values only at certain values of x, such as evaluated by sampling or by any other mean.
- Some functions do not possess the closed form solutions like $I_1=\int_0^x e^{-v^2}dv,\ I_2=\int_0^x \frac{u^3}{e^u-1}du$ etc.
- Some functions possess their anti-derivative form but in complex form and also their numerical approximation can be easily calculated.

A definite integral $I(x)=\int_a^b f(x)\,dx$; is to be evaluated numerically in numerical integration. Since, it can be treated as the area under the curve f(x), enclosed between the limits a and b. This process of computation is called as Quadrature. Now, the main goal is to compute the integral I(x) approximately by calculating the area beneath the curve y = f(x). For calculating this area, we may use the virtual strips as shown in the figure:

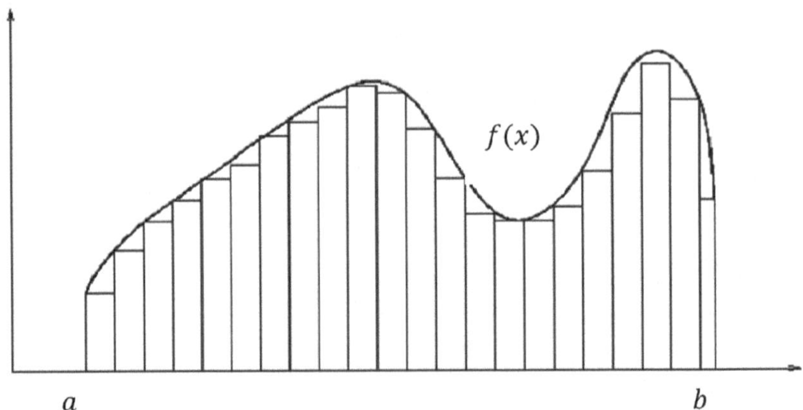

Fig. 7.1: Strips under the curve

From figure 7.1, it is clear that we can find the better approximation of area by fining the widths of rectangular strips.

Let we have to find the approximate value of the integral by evaluating f(x) at the finite number of sample points i.e.

$a = x_0 < x_1 < x_2 < x_3 < ..... < x_n = b$

and assumed that

$$\int_a^b f(x)\,dx = Q_n(f) + E_n(f) \tag{7.1.1}$$

where; Q_n (f) is approximate value of integral which would be evaluated numerically and E_n (f) is truncation error.

To evaluate $\int_a^b f(x)\,dx$ numerically, equation (7.1.1) is called as Quadrature formula. There are various methods to choose the location and number of sampling points $\{x_i\}$. For Newton-Cotes methods, x_i's are chosen to be equally spaced and for Gauss-Legendre Quadrature, are chosen as zeros of certain Legendre's polynomials.

7.1.1 Relative Error

Recall equation (7.1.1):

$$\int_a^b f(x)\,dx = Q_n(f) + E_n(f)$$

where; $Q_n(f)$ is approximate value of integral which would be evaluated numerically and $E_n(f)$ is truncation error which is equal to:

$$E_n(f) = \int_a^b f(x)\,dx - Q_n(f),$$

$$E_n(f) = \text{exact value of integral} - \text{calculated value of integral}$$

and relative error is defined as:

$$E_r(f) = \frac{\text{exact value of integral} - \text{calculated value of integral}}{\text{exact value of integral}}$$

7.2 Newton-Cotes Closed Quadrature Formula

Newton-Cotes Closed Quadrature Formulae are the basis for number of numerical integration methods. Also, these are most popular and widely used numerical integration formulae. Their derivation is based on Newton or Lagrange's polynomial interpolation. Generally, Quadrature formula (7.1.1) is taken as:

$$Q_n(f) = \sum_{i=0}^{n} w_i\, f(x_i);$$

where; x_i`s are sampling points or called as integration nodes.

and w_i`s are called as weighting coefficients or weights.

Therefore, equation (7.1.1) can be written as:

$$\int_a^b f(x)\,dx \approx \sum_{i=0}^{n} w_i\, f(x_i)$$

In these formula, It is considered that there exist a unique polynomial $P_n(x)$ of degree $\leq$ n which pass through n + 1 equally spaced sampling points $(x_i, f(x_i))$ where and this polynomial is used to approximate f(x) over [a,b] i.e.

$$\int_a^b f(x)\,dx \approx \int_a^b P_n(x)\,dx$$

Thus; the resulting formulae are called as Newton-Cotes Quadrature Formulae. If the limits of integration (a and b) are in set $\{x_i\}$, then Quadrature formula is called as closed and if limits lie beyond the set $\{x_i\}$, then Quadrature formula is called as open. But open form formula cannot be used in definite integral. Thus, the resulting Quadrature formulae are called as Newton-Cotes closed Quadrature Formulae. They include:

- Trapezoidal rule (Two-point form, n = 1)

$$\int_{x_0}^{x_1} f(x)\,dx \approx \frac{h}{2}(f_0 + f_1)$$

- Simpson's $\frac{1}{3}$ rule (Three-point form, n = 2)

$$\int_{x_0}^{x_2} f(x)\,dx \approx \frac{h}{3}(f_0 + 4f_1 + f_2)$$

- Simpson's $\frac{3}{8}$ rule (Four-point form, n = 3)

$$\int_{x_0}^{x_3} f(x)\,dx \approx \frac{3h}{8}(f_0 + 3f_1 + 3f_2 + f_3)$$

- Boole's rule (Five-point form, n = 4)

$$\int_{x_0}^{x_4} f(x)\,dx \approx \frac{14h}{45}(7f_0 + 32f_1 + 12f_2 + 32f_3 + 7f_4)$$

- Weddle's rule (Six-point form, n = 5)

$$\int_{x_0}^{x_5} f(x)\,dx \approx \frac{3h}{10}(f_0 + 5f_1 + f_2 + 6f_3 + f_4 + 5f_5 + f_6)$$

where $f_i = f(x_i)$ and $h = x_i - x_{i-1}$; $i = 1,2,\ldots,n$.

In their derivations, polynomial $P_n(X)$ is used which is interpolated by using either Lagrange's interpolation formula or Newton's forward difference formula. These interpolated formulae are given as follows:

According to Lagrange's interpolation formula:

$$P_n(x) = \frac{(x-x_1)(x-x_2)\ldots(x-x_n)}{(x_0-x_1)(x_0-x_2)\ldots(x_0-x_n)} f_0 + \frac{(x-x_0)(x-x_2)\ldots(x-x_n)}{(x_1-x_0)(x_1-x_2)\ldots(x_1-x_n)} f_1 + \ldots$$
$$+ \frac{(x-x_0)(x-x_1)\ldots(x-x_{n-1})}{(x_n-x_0)(x_n-x_1)\ldots(x_n-x_{n-1})} f_n, \tag{7.2.1}$$

and according to Newton's forward difference formula:

$$P_n(p) = f_0 + \frac{p}{1!}\Delta f_0 + \frac{p(p-1)}{2!}\Delta^2 f_0 + \ldots + \frac{p(p-1)\ldots(p-n+1)}{n!}\Delta^n f_0 \qquad (7.2.2)$$

where, $p = \dfrac{x - x_0}{h}$

7.3 Gauss-Legendre Quadrature Formulae

In Newton-Cotes Quadrature Formulae, the nodes of integration are taken equally spaced but sometimes it would be more convenient to use unequal intervals for $\{X_i\}$. In these rules, f(X) is approximated in the following way:

$$\int_a^b f(x)\,dx = \int_a^b w(x)\ f(x)\,dx = w_0\ f(x_0) + w_1\ f(x_1) + \ldots + w_n\ f(x_n)$$

where w(x) is weight function and x_i's are taken as zeros of an orthogonal polynomial; orthogonal with respect to weight function w(x) over [a, b] and the weights are given by

$$w_i = \int_a^b w(x)\ L_i(x)\,dx$$

where $L_i(x)$ are Legendre's fundamental polynomials and also $L_i(x) > 0$.

Generally, limits of integration are taken as -1 and 1 in these rules. If there are no such limits then we can reduce the interval [a, b] to [– 1,1] to by using a simple linear transformation as:

$x = \dfrac{(b-a)t + (b+a)}{2}$; then we have:

$$\int_a^b f(x)\,dx = \frac{b-a}{2}\int_{-1}^{1} f\left(\frac{(b-a)t + (b+a)}{2}\right)dt = \int_{-1}^{1} g(t)\,dt$$

where $g(t) = \dfrac{b-a}{2} f\left(\dfrac{(b-a)t + (b+a)}{2}\right)$

In these rules, the set of nodes of integration contains one or more than one node. On the basis of number of integration nodes, these rules can be classified in the following way:

- **Gauss-Legendre one point rule**

The Gauss-Legendre's one point rule for calculating the approximate value of the integral $\int_{-1}^{1} f(x)\,dx$ is defined as:

$$\int_{-1}^{1} f(x)dx = w_0 f(x_0);\ w_0 \neq 0 \tag{7.3.1}$$

equation (7.3.1) has two parameters w_o, x_o. Therefore, degree of *f(x)* should be ≤ 2 and (7.3.1) is exact for linear polynomial $f(x) = a + bx$. Since two parameters w_o, x_o are to be determined therefore we can select two conditions to be satisfied. Let these are defined as:

Now, for $f(x) = 1$;

$$\int_{-1}^{1} 1\, dx = 2 = w_o$$

Now, for $f(x) = x$;

$$\int_{-1}^{1} x\, dx = 0 = w_o x_o$$

Since, $w_o \neq 0$, $\therefore\ x_o = 0$; thus, equation (7.3.1) becomes as:

$$\int_{-1}^{1} f(x)dx = 2f(0); \tag{7.3.2}$$

Hence, we observe that if we have taken one node of integration then it would be zero and equation (7.3.2) is Gauss-Legendre's one point Quadrature formula for calculating the approximate value of integral $\int_{-1}^{1} f(x)dx$ numerically.

- **Gauss-Legendre two point rule**

The Gauss-Legendre's two point rule for calculating the approximate value of the integral $\int_{-1}^{1} f(x)dx$ is defined as:

$$\int_{-1}^{1} f(x)dx = w_0 f(x_0) + w_1 f(x_1);\ \ w_0, w_1 \neq 0 \tag{7.3.3}$$

equation (7.3.3) has four parameters w_0, x_0, w_1, x_1. Therefore, degree of f(x) should be ≤ 4 and to make (7.3.3) exact; we can select f(x) as $f(x) = 1, x, x^2, x^3$.

Now, for $f(x) = 1$

$$\int_{-1}^{1} 1\, dx = 2 = w_o(1) + w_1(1)$$

for f(x) = x;

$$\int_{-1}^{1} x\,dx = 0 = w_o x_o + w_1 x_1$$

for f(x) = x^2;

$$\int_{-1}^{1} x^2\,dx = \frac{2}{3} = w_o x_o^{\,2} + w_1 x_1^{\,2}$$

for f(x) = x^3;

$$\int_{-1}^{1} x^3\,dx = 0 = w_o x_o^{\,3} + w_1 x_1^{\,3}$$

and solution of above equations is:

$$w_o = w_1 = 1,\ \ x_o = \frac{1}{\sqrt{3}} \approx 0.5773502692,\ \ x_1 = -\frac{1}{\sqrt{3}} \approx -0.5773502692;$$

Hence, equation (7.3.3) becomes as:

$$\int_{-1}^{1} f(x)\,dx = f\left(\frac{1}{\sqrt{3}}\right) + f\left(-\frac{1}{\sqrt{3}}\right);$$

$$\approx f(0.5773502692) + f(-0.5773502692) \qquad (7.3.4)$$

Equation (7.3.4) is Gauss-Legendre's two point Quadrature formula for calculating the approximate value of integral $\int_{-1}^{1} f(x)\,dx$ numerically.

Similarly, we can extend these rules by evaluating weights and nodes for three points, four points and so on. All these parameters are listed in the table 7.1:

Table 7. 1: Weights and nodes used in Gauss-Legendre's Quadrature formulae

Number of points	x_i	Approximate value of x_i	Weights, w_i	Approximate value of w_i
1	0	0	2	2
2	$\pm\frac{1}{\sqrt{3}}$	± 0.57735	1	1
3	0	0	$\frac{8}{9}$	0.888889
	$\pm\sqrt{\frac{3}{5}}$	± 0.774597	$\frac{5}{9}$	0.555556
4	$\pm\sqrt{\frac{3}{7}-\frac{2}{7}\sqrt{\frac{6}{5}}}$	± 0.339981	$\frac{18+\sqrt{30}}{36}$	0.652145
	$\pm\sqrt{\frac{3}{7}+\frac{2}{7}\sqrt{\frac{6}{5}}}$	± 0.861136	$\frac{18-\sqrt{30}}{36}$	0.347855
5	0	0	$\frac{128}{225}$	0.568889
	$\pm\frac{1}{3}\sqrt{5-2\sqrt{\frac{10}{7}}}$	± 0.538469	$\frac{322+13\sqrt{70}}{900}$	0.478629
	$\pm\frac{1}{3}\sqrt{5+2\sqrt{\frac{10}{7}}}$	± 0.90618	$\frac{322-13\sqrt{70}}{900}$	0.236927

Example 7.1: Find the Legendre's polynomial of degree 2.

Solution: Legendre's polynomial of degree s is defined as:

$$P_s(x)=\frac{1}{2^s\ s!}\frac{d^s}{dx^s}\left(x^2-1\right)^s$$

Therefore,

$$P_2(x)=\frac{1}{2^2\,2!}\frac{d^2}{dx^2}\left(x^2-1\right)^2,$$

$$P_2(x)=\frac{1}{4\times 2}\frac{d}{dx}\left(2\times\left(x^2-1\right)\times 2x\right),$$

$$P_2(x)=\frac{1}{8}\frac{d}{dx}\left(4x^3-4x\right),$$

$$P_2(x) = \frac{1}{8}\left(12x^2 - 4\right) = \frac{1}{2}\left(3x^2 - 1\right),$$

Example 7.2: Use Gauss-Legendre's four point formula to find approximate value of

$$\int_0^{10} e^{\frac{-1}{1+x^2}} dx;$$

Solution: By using Gauss-Legendre's four point formula, we have:

$$\int_{-1}^{1} f(x)dx = w_0 f(x_0) + w_1 f(x_1) + w_2 f(x_2) + w_3 f(x_3);$$

Now,

$$x_0 = -0.86114,\ x_1 = -0.33998,\ x_2 = 0.33998,\ x_3 = 0.86114,$$

$$w_0 = 0.34785,\ w_1 = 0.65215,\ w_2 = 0.65215,\ w_3 = 0.34785,$$

$$f(x) = e^{\frac{-1}{1+x^2}}$$

$$\Rightarrow f_0 = f(-0.86114) = e^{\frac{-1}{1+(-0.86114)^2}} = 0.56316,$$

$$\Rightarrow f_1 = f(-0.33998) = e^{\frac{-1}{1+(-0.33998)^2}} = 0.40804,$$

$$\Rightarrow f_2 = f(0.33998) = e^{\frac{-1}{1+(0.33998)^2}} = 0.40804,$$

$$\Rightarrow f_3 = f(0.86114) = e^{\frac{-1}{1+(-0.86114)^2}} = 0.56316,$$

$$\int_{-1}^{1} e^{\frac{-1}{1+x^2}} dx = w_0 f(x_0) + w_1 f(x_1) + w_2 f(x_2) + w_3 f(x_3)$$

$$= 0.34785 \times 0.56316 + 0.65215 \times 0.40804 + 0.65215 \times 0.40804 + 0.34785 \times 0.56316$$

$$= 0.924$$

Example 7.3: Use Gauss-Legendre's five point formula to find approximate value of

$$\int_0^3 5\,x^4\,dx;$$

Solution: By using Gauss-Legendre's five point formula, we have:

$$\int_{-1}^{1} f(x)\,dx = w_0 f(x_0) + w_1 f(x_1) + w_2 f(x_2) + w_3 f(x_3) + w_4 f(x_4);$$

First of all, we change the limits of integration from [0,3] to [-1,1]. Therefore, put

$$t = \frac{0+3}{2} + \frac{3-0}{2}x = 1.5 + 1.5x$$

$$\Rightarrow dt = 1.5\ dx$$

$$\Rightarrow \int_0^3 5\,x^4\,dx = \frac{5}{1.5}\int_{-1}^{1}\left(\frac{t-1.5}{1.5}\right)^4 dt = \frac{5}{(1.5)^5}\int_{-1}^{1}(t-1.5)^4\,dt$$

Now, we first find the approximate value of integral $\int_{-1}^{1}(t-1.5)^4\,dt$ as follows:

$t_0 = -0.90618$, $t_1 = -0.53847$, $t_2 = 0$, $t_3 = 0.53849$, $t_4 = 0.90618$,

$w_0 = 0.23693$, $w_1 = 0.47863$, $w_2 = 0.56889$, $w_3 = 0.47863$, $w_4 = 0.23693$,

$g(t) = (t - 1.5)^4$

$\Rightarrow g_0 = g(-0.90618) = (-0.90618 - 1.5)^4 = 33.5207$,

$\Rightarrow g_1 = g(-0.53847) = (-0.53847 - 1.5)^4 = 17.2670$,

$\Rightarrow g_2 = g(0) = (0 - 1.5)^4 = 5.0625$,

$\Rightarrow g_3 = g(0.53849) = (0.53849 - 1.5)^4 = 0.85477$,

$\Rightarrow g_4 = g(0.90618) = (0.90618 - 1.5)^4 = 0.12434$,

$$\int_{-1}^{1}(t-1.5)^4\,dt = w_0 g(x_0) + w_1 g(x_1) + w_2 g(x_2) + w_3 g(x_3) + w_4 g(x_4)$$

$= 0.23693 \times 33.5207 + 0.47863 \times 17.2670 + 0.56889 \times 5.0625 + 0.47863 \times 0.85477 + 0.23693 \times 0.12434$

$= 19.5251$

Therefore; we have:

$$\int_0^3 5x^4\,dx = \frac{5}{(1.5)^5}\int_{-1}^{1}(t-1.5)^4\,dt = \frac{5}{(1.5)^5}\times 19.5251 = 19.28405$$

7.4 Gaussian Integration

Carl Friedrich Guass presented a Quadrature Rule to yield an exact result for polynomials of degree 2n – 1 or less by an approximate choice of integration nodes x_i and weights w_i. In this rule, the integral $\int_a^b f(x)\,dx$ is approximated as follows:

$$\int_a^b f(x)\,dx = \sum_{i=1}^{s} w_i\, f(x_i)$$

Gaussian Integration Quadrature rules are useful whether limits are finite or one of the limits is infinite or both the limits are infinite. These rules depend on limits of integration and expression for weight functions in the following way:

Table 7. 2: List of Gaussian Integration Quadrature rules

Type of Gaussian Integration Quadrature rules	Limits of integration	Nodes of integration	Weights
Gauss-Legendre Quadrature Rules	[-1,1]	Zeros of corresponding Legendre's polynomial	
Gauss-Chebyshev Quadrature Rules	[-1,1]	Zeros of corresponding Chebyshev's polynomial	$\frac{1}{\sqrt{1-x^2}}$
Gauss-Laguerre Quadrature Rules	[0, ∞]	Zeros of corresponding Laguerre's polynomial	e^x
Gauss-Hermite Quadrature Rules	(0, ∞)	Zeros of corresponding Hermite's polynomial	e^{-x^2}

7.5 Euler-Meclaurin Formula

Euler Maclaurin Quadrature formula is a link between the integral and sums.

Let, we have equi-spaced points $x_1, x_2, \ldots, x_n$ and $\int_{x_0}^{x_n} f(x)\,dx$ is to be evaluated numerically. Consider the function F(x) such that $f(x)= \Delta F(x)$ and we have:

$\Rightarrow F(x) = \Delta^{-1} f(x),$

$\Rightarrow F = \Delta^{-1} f,$

$\Rightarrow F = (E+1)^{-1} f,$

$\Rightarrow F=(e^{hD}-1)^{-1} f,$

$$\Rightarrow F=\left(\left(1+\frac{hD}{1!}+\frac{(hD)^2}{2!}+\frac{(hD)^3}{3!}+\ldots\right)-1\right)^{-1} f,$$

$$\Rightarrow F=\left(hD\left(1+\frac{(hD)^1}{2!}+\frac{(hD)^2}{3!}+\ldots\right)\right)^{-1} f,$$

$$\Rightarrow F=(hD)^{-1}\left(1+\frac{(hD)^1}{2!}+\frac{(hD)^2}{3!}+\ldots\right)^{-1} f,$$

$$\Rightarrow F=\frac{1}{h}D^{-1}\left(1-\frac{(hD)^1}{2}+\frac{(hD)^2}{12}-\frac{(hD)^4}{720}+\ldots\right)f,$$

$$\Rightarrow F=\frac{1}{h}\left(D^{-1}-\frac{h}{2}+\frac{h^2D}{12}-\frac{h^4D^3}{720}+\ldots\right)f,$$

$$\Rightarrow F(x)=\frac{1}{h}\left(\int f(x)\,dx-\frac{h}{2}f(x)+\frac{h^2}{12}f'(x)-\frac{h^4}{720}f'''(x)+\ldots\right) \quad (7.5.1)$$

Put $x = x_n$ and $x = x_0$ in equation (7.5.1) then subtract the resulting equation from (7.5.1); we got:

$\Rightarrow F(x_n) - F(x_0)$

$$=\frac{1}{h}\left(\int_{x_0}^{x_n} f(x)\,dx-\frac{h}{2}\left(f(x_n)-f(x_0)\right)+\frac{h^2}{12}\left(f'(x_n)-f'(x_0)\right)-\frac{h^4}{720}\left(f'''(x_n)-f'''(x_0)\right)+\ldots\right)$$

(7.5.2)

Now, f(x)=ΔF(x),

$\Rightarrow f(x_0) = F(x_1) - F(x_0)$,

$\Rightarrow f(x_1) = F(x_2) - F(x_1)$,

....

$\Rightarrow f(x_{n-1}) = F(x_n) - F(x_{n-1})$,

$\Rightarrow f(x_0) + f(x_1) + \ldots\ldots + f(x_{n-1}) = F(x_n) - F(x_0)$,

$$\Rightarrow \sum_{i=0}^{n-1} f(x) = F(x_n) - F(x_0), \qquad (7.5.3),$$

from equations (7.5.2) and (7.5.3), we have:

$$\Rightarrow \sum_{i=0}^{n-1} f(x) = \frac{1}{h}\left(\int_{x_0}^{x_n} f(x)\,dx - \frac{h}{2}\left(f(x_n) - f(x_0)\right) + \frac{h^2}{12}\left(f'(x_n) - f'(x_0)\right) - \frac{h^4}{720}\left(f'''(x_n) - f'''(x_0)\right) + \ldots\right),$$

$$\Rightarrow \sum_{i=0}^{n-1} f(x) = \frac{1}{h}$$
$$\left(\int_{x_0}^{x_n} f(x)\,dx - \frac{h}{2}\left(f(x_n) - f(x_0)\right) + \frac{h^2}{12}\left(f'(x_n) - f'(x_0)\right) - \frac{h^4}{720}\left(f'''(x_n) - f'''(x_0)\right) + \ldots\right)$$

$$\Rightarrow \sum_{i=0}^{n-1} f(x) = \frac{1}{h}\int_{x_0}^{x_n} f(x)\,dx - \frac{1}{2}\left(f(x_n) - f(x_0)\right) + \frac{h}{12}\left(f'(x_n) - f'(x_0)\right)$$
$$-\frac{h^3}{720}\left(f'''(x_n) - f'''(x_0)\right) + \ldots,$$

$$\Rightarrow \frac{1}{h}\int_{x_0}^{x_n} f(x)\,dx$$
$$= \sum_{i=0}^{n-1} f(x) + \frac{1}{2}\left(f(x_n) - f(x_0)\right) - \frac{h}{12}\left(f'(x_n) - f'(x_0)\right)$$
$$+\frac{h^3}{720}\left(f'''(x_n) - f'''(x_0)\right) - \ldots\ldots$$

$$\Rightarrow \int_{x_0}^{x_n} f(x)\,dx = \frac{h}{2}\left(f(x_0) + 2f(x_1) + 2f(x_2) + \ldots + 2f(x_{n-1}) + f(x_n)\right) - \frac{h^2}{12}\left(f'(x_n) - f'(x_0)\right)$$
$$+\frac{h^4}{720}\left(f'''(x_n) - f'''(x_0)\right) - \ldots, \qquad (7.5.4)$$

Equation (7.5.4) is called as Euler-Maclaurin Quadrature formula for evaluating the value of integrand f(x) numerically.

Example 7.4: Find the approximate value of the following series:

$$\frac{1}{1^2}+\frac{1}{3^2}+\frac{1}{5^2}+\ldots+\frac{1}{49^2}$$

Solution: According to Euler-Maclaurin formula, we have:

$$\Rightarrow \sum_{i=0}^{n-1} f(x)=\frac{1}{h}$$

$$\int_{x_0}^{x_n} f(x)\,dx-\frac{1}{2}\left(f(x_n)-f(x_0)\right)+\frac{h}{12}\left(f'(x_n)-f'(x_0)\right)-\frac{h^3}{720}\left(f'''(x_n)-f'''(x_0)\right)+\ldots,$$

$$\text{Now, } f(x)=\frac{1}{x^2},\; f'(x)=\frac{-2}{x^3}, f'''(x)=\frac{-24}{x^5},\quad x_0=1,\, x_n=49,\;\; h=2,\, n=25$$

$$\Rightarrow \sum_{i=0}^{n-1} f(x)=\frac{1}{2}\int_1^{49}\frac{1}{x^2}\,dx-\frac{1}{2}\left(\frac{1}{49^2}-\frac{1}{1^2}\right)+\frac{2}{12}\left(\frac{-2}{49^3}-\frac{-2}{1^3}\right)-\frac{2^3}{720}\left(\frac{-24}{49^5}-\frac{-24}{1^5}\right)+\ldots,$$

$$\Rightarrow \sum_{i=0}^{n-1} f(x)=0.4898+0.4998+0.3333-0.2667+\ldots,$$

$$\Rightarrow \sum_{i=0}^{n-1} f(x)\approx 1.0562,$$

7.6 Trapezoidal Rule

It is the Newton-Cote's closed Quadrature formula in which two nodes are taken i.e. $\{x_0, x_1\}$ and n = 1. Here, first order interpolation polynomial P_1 (x) is used to approximate the integrand f(x) over these two points.

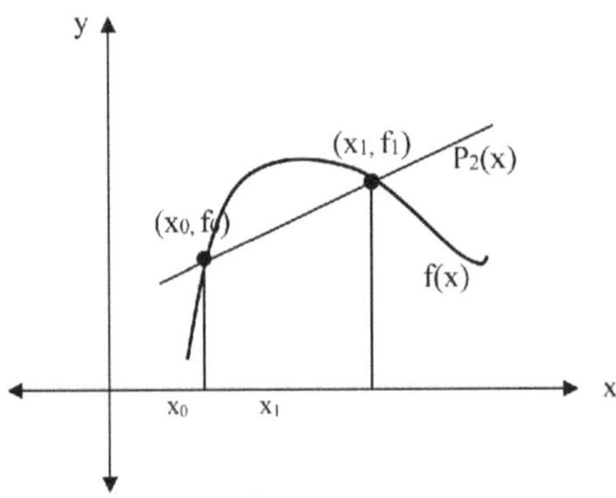

Fig. 7.2: Approximation of curve by using a trapezoid

As shown in figure, f(x) is approximated by a trapezoid with horizontal base width h ($h = x_1 - x_0$) and sloped top [$f_0 = f(x_0)$, $f_1 = f(x_1)$]. A polynomial function $P_1(x)$ is used for this approximation which can be defined in the following two ways:

By using Lagrange's Interpolation Formula

Now, Lagrange's interpolating polynomial (7.2.1) throughout the points (x_0, $f(x_0)$) and (x_1, $f(x_1)$) is defined as follows:

$$P_1(x) = \frac{x - x_1}{x_0 - x_1} f_0 + \frac{x - x_0}{x_1 - x_0} f_1$$

$$= \frac{x - x_1}{-h} f_0 + \frac{x - x_0}{h} f_1$$

$$= \frac{1}{h}\left[-(x - x_1) f_0 + (x - x_0) f_1\right]$$

$$= \frac{1}{h}\left[x(f_1 - f_0) + x_1 f_0 - x_0 f_1\right]$$

Now,

$$\int_{x_0}^{x_1} f(x)\,dx = \int_{x_0}^{x_1} P_1(x)\,dx$$

$$= \int_{x_0}^{x_1} \left(\frac{1}{h}\left[x(f_1 - f_0) + x_1 f_0 - x_0 f_1\right]\right) dx$$

$$= \frac{1}{h}(f_1 - f_0)\int_{x_0}^{x_1} (x)\,dx + \left(\frac{1}{h}[x_1 f_0 - x_0 f_1]\right)\int_{x_0}^{x_1} 1\,dx$$

$$= \frac{1}{h}(f_1 - f_0)\int_{x_0}^{x_1} (x)\,dx + \left(\frac{1}{h}[x_1 f_0 - x_0 f_1]\right)\int_{x_0}^{x_1} 1\,dx$$

$$= \frac{1}{h}(f_1 - f_0)\left|\frac{x^2}{2}\right|_{x_0}^{x_1} + \left(\frac{1}{h}[x_1 f_0 - x_0 f_1]\right)|x|_{x_0}^{x_1}$$

$$=\frac{1}{h}(f_1-f_0)\left(\frac{x_1^{\ 2}-x_0^{\ 2}}{2}\right)+\left(\frac{1}{h}[x_1f_0-x_0f_1]\right)(x_1-x_0)$$

$$=\frac{1}{h}(f_1-f_0)\left(\frac{(x_1-x_0)(x_1+x_0)}{2}\right)+\left(\frac{1}{h}[x_1f_0-x_0f_1]\right)(x_1-x_0)$$

$$=\frac{(x_1-x_0)}{h}\left[(f_1-f_0)\left(\frac{(x_1+x_0)}{2}\right)+(x_1f_0-x_0f_1)\right]$$

$$=\frac{h}{h}\left[(f_1-f_0)\left(\frac{(x_1+x_0)}{2}\right)+(x_1f_0-x_0f_1)\right]$$

$$=\frac{1}{2}[(f_1-f_0)(x_1+x_0)+2(x_1f_0-x_0f_1)]$$

$$=\frac{1}{2}[(f_1-f_0)(x_0+h+x_0)+2((x_0+h)f_0-x_0f_1)]$$

$$=\frac{1}{2}[(f_1-f_0)(2x_0+h)+2(x_0f_0+hf_0-x_0f_1)]$$

$$=\frac{1}{2}[2x_0f_1+hf_1-2x_0f_0-hf_0+2x_0f_0+2hf_0-2x_0f_1]$$

$$=\frac{1}{2}[hf_1+hf_0]$$

$$=\frac{h}{2}[f_1+f_0]$$

By using Newton's forward difference formula:

According to Newton's forward difference formula (7.2.2), we can take P_1 1 (*x*) as follows:

$P_1(p) = f_0 + p\Delta f_0$

Now, $\int_{x_0}^{x_1} f(x)dx = \int_{x_0}^{x_1} P_1(x)dx$

$$= h\int_0^1 (f_0 + p\Delta f_0)\, dp$$

$$= h\left(f_0 \left|p\right|_0^1 + \Delta f_0 \left|\frac{p^2}{2}\right|_0^1 \right)$$

$$= h\left(f_0 + \frac{1}{2}\Delta f_0 \right)$$

$$= h\left(f_0 + \frac{1}{2}(f_1 - f_0) \right)$$

$$= h\left(\frac{f_1 + f_0}{2} \right)$$

$$\therefore \int_{x_0}^{x_1} f(x)dx = \frac{h}{2}[f_1 + f_0] \tag{7.6.1}$$

7.6.1 Error Estimate in Trapezoidal Rule

In trapezoidal rule, to find the integration of integrand f(x), we have approximated it by using linear interpolating function. We also considered that f(x) is sufficiently differentiable. Therefore, an error $E_{TR}(f(x))$ for this Newton-Cotes Quadrature formula has contained an approximate higher derivative. Therefore, by using equation (7.2.2), we have the higher derivative term as:

$$h^2 f''(c_n)\int_0^1 \frac{p(p-1)}{2!}\, h\, dp = h^3 \frac{f''(c_n)}{2}\left|\frac{p^3}{3} - \frac{p^2}{2}\right|_0^1 = -\frac{h^3}{12} f''(c_n)$$

where $a < c_n < b$

Thus, equation (7.6.1) can be written as:

$$\int_{x_0}^{x_1} f(x)dx = \frac{h}{2}[f_1 + f_0] - \frac{h^3}{12} f''(c_n),$$

where,

$$E_{TR}(f(x)) = -\frac{h^3}{12} f''(c_n),$$

Example 7.5: Evaluate the integral $\int_0^2 (4x^2 + 3x - 5)dx$ by using trapezoidal rule.

Solution: According to trapezoidal rule, we have:

$$\int_{x_0}^{x_1} f(x)dx = \frac{h}{2}[f_1 + f_0]$$

Now,

$x_0 = 0,\ x_1 = 2,\ h = x_1 - x_0 = 2 - 0 = 2$

and $f(x) = 4x^2 + 3x - 5$

$\Rightarrow f_0 = f(0) = 4(0^2) + 3\ (0) - 5 = -5,$

$\Rightarrow f_1 = f(2) = 4(2^2) + 3\ (2) - 5 = 17;$

$$\therefore \int_{x_0}^{x_1} f(x)dx = \frac{h}{2}[f_1 + f_0] = \frac{2}{2}[17 - 5] = 12$$

Hence, the approximate value of the integral is 12.

7.7 Simpson's 1/3 Rule

It is the Newton-Cote's closed Quadrature formula in which three nodes are taken i.e. $\{x_0, x_1, x_2\}$ and n=2. Here, second order interpolation polynomial $P_2 2\ (x)$ is used to approximate the integrand f(x) over these points.

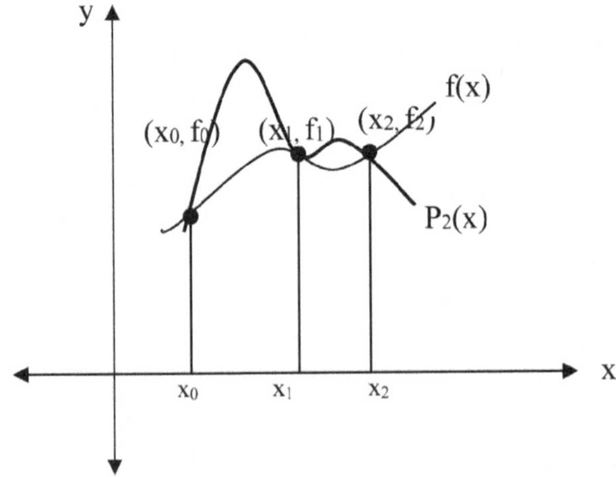

Fig. 7.3: Approximation of area by using two trapezoids

As shown in figure, f(x) is approximated by the combinations of trapezoids with horizontal base width h ($h = x_i - x_{i-1}$); i = 1,2 and sloped tops [$f_i = f(x_i)$, $f_{i-1} = f(x_{i-1})$]. Now, an interpolating polynomial throughout the points (x_0, $f(x_0)$), (x_1, $f(x_1)$) and (x_2, $f(x_2)$) is used for this approximation which can be defined in the following two ways:

By using Lagrange's Interpolation Formula:

According to Lagrange's interpolation formula (7.2.1), we have:

$$P_2(x) = \frac{(x-x_1)(x-x_2)}{(x_0-x_1)(x_0-x_2)}f_0 + \frac{(x-x_0)(x-x_2)}{(x_1-x_0)(x_1-x_2)}f_1 + \frac{(x-x_0)(x-x_1)}{(x_2-x_0)(x_2-x_1)}f_2$$

$$= \frac{(x-x_1)(x-x_2)}{2h^2}f_0 - \frac{(x-x_0)(x-x_2)}{2h^2}f_1 + \frac{(x-x_0)(x-x_1)}{2h^2}f_2$$

$$= \frac{1}{2h^2}$$

$$\left[x^2(f_0 - 2f_1 + f_2) + x(-x_1f_0 - x_2f_0 + 2x_2f_1 + 2x_0f_1 - x_0f_2 - x_1f_2) + (x_1x_2f_0 - x_0x_2f_1 + x_0x_1f_2)\right]$$

Now,

$$\int_{x_0}^{x_2} f(x)\,dx = \int_{x_0}^{x_2} P_2(x)\,dx$$

$$= \int_{x_0}^{x_2}\left(\frac{1}{2h^2}\left[x^2(f_0 - 2f_1 + f_2) + x(-x_1f_0 - x_2f_0 + 2x_2f_1 + 2x_0f_1 - x_0f_2 - x_1f_2) + (x_1x_2f_0 - x_0x_2f_1 + x_0x_1f_2)\right]\right)$$

$$= \frac{1}{2h^2}$$

$$\left[\int_{x_0}^{x_2}\left(x^2(f_0 - 2f_1 + f_2)\right)dx + \int_{x_0}^{x_2}\left(x(-x_1f_0 - x_2f_0 + 2x_2f_1 + 2x_0f_1 - x_0f_2 - x_1f_2)\right)dx + \int_{x_0}^{x_2}(x_1x_2f_0 - x_0x_2f_1 + x_0x_1f_2)\,dx\right]$$

$$= \frac{1}{2h^2}$$

$$\left[(f_0 - 2f_1 + f_2)\int_{x_0}^{x_2}(x^2)\,dx + (-x_1f_0 - x_2f_0 + 2x_2f_1 + 2x_0f_1 - x_0f_2 - x_1f_2)\int_{x_0}^{x_2}(x)\,dx + (x_1x_2f_0 - x_0x_2f_1 + x_0x_1f_2)\int_{x_0}^{x_2}1\,dx\right]$$

$$= \frac{1}{2h^2}$$

$$\left[(f_0 - 2f_1 + f_2)\left|\frac{x^3}{3}\right|_{x_0}^{x_2} + (-x_1f_0 - x_2f_0 + 2x_2f_1 + 2x_0f_1 - x_0f_2 - x_1f_2)\left|\frac{x^2}{2}\right|_{x_0}^{x_2} + (x_1x_2f_0 - x_0x_2f_1 + x_0x_1f_2)|x|_{x_0}^{x_2}\right]$$

$$=\frac{1}{2h^2}\left[\begin{array}{l}(f_0-2f_1+f_2)\left(\frac{x_2^{\ 3}-x_0^{\ 3}}{3}\right)+(-x_1f_0-x_2f_0+2x_2f_1+2x_0f_1-x_0f_2-x_1f_2)\\ \left(\frac{x_2^{\ 2}-x_0^{\ 2}}{2}\right)+(x_1x_2f_0-x_0x_2f_1+x_0x_1f_2)(x_2-x_0)\end{array}\right]$$

$$=\frac{1}{2h^2}\left[\begin{array}{l}(f_0-2f_1+f_2)\left(\frac{(x_2-x_0)(x_2^{\ 2}+x_0^{\ 2}-x_2x_0)}{3}\right)+(-x_1f_0-x_2f_0+2x_2f_1+2x_0f_1-x_0f_2-x_1f_2)\\ \left(\frac{(x_2-x_0)(x_2+x_0)}{2}\right)+(x_1x_2f_0-x_0x_2f_1+x_0x_1f_2)(x_2-x_0)\end{array}\right]$$

$$=\frac{(x_2-x_0)}{2h^2}\left[\begin{array}{l}(f_0-2f_1+f_2)\left(\frac{(x_2^{\ 2}+x_0^{\ 2}-x_2x_0)}{3}\right)+(-x_1f_0-x_2f_0+2x_2f_1+2x_0f_1-x_0f_2-x_1f_2)\\ \left(\frac{(x_2+x_0)}{2}\right)+(x_1x_2f_0-x_0x_2f_1+x_0x_1f_2)\end{array}\right]$$

$$=\frac{2h}{2h^2}\left[\begin{array}{l}(f_0-2f_1+f_2)\left(\frac{(x_2^{\ 2}+x_0^{\ 2}-x_2x_0)}{3}\right)+(-x_1f_0-x_2f_0+2x_2f_1+2x_0f_1-x_0f_2-x_1f_2)\\ \left(\frac{(x_2+x_0)}{2}\right)+(x_1x_2f_0-x_0x_2f_1+x_0x_1f_2)\end{array}\right]$$

On simplifying; we got:

$$\int_{x_0}^{x_2} f(x)dx=\frac{h}{3}[f_0+4f_1+f_2]$$

Alternate method of Simplification:

We have integration nodes as $\{x_0, x_1, x_2\}$ with width of strip as h and we can use the linear transformation as: $x = x_0 + ht \rightarrow dx = dt$ and $t \rightarrow 0$ as $x \rightarrow x_0$; $t \rightarrow 2$ as $x \rightarrow x_2$

also, $x - x_0 = ht$; $x - x_1 = h(t-1)$; $x-x_2 = h\ (t-2)$;

and $(x_2 - x_0) = 2h$; $(x_2 - x_1) = h$; $(x_1 - x_0) = h$

$$\therefore P_2(x)=\frac{(x-x_1)(x-x_2)}{(x_0-x_1)(x_0-x_2)}f_0+\frac{(x-x_0)(x-x_2)}{(x_1-x_0)(x_1-x_2)}f_1+\frac{(x-x_0)(x-x_1)}{(x_2-x_0)(x_2-x_1)}f_2$$

becomes as follows:

$$P_2(x)=\frac{h(t-1)h(t-2)}{(-h)(-2h)}f_0+\frac{h(t)h(t-2)}{(h)(-h)}f_1+\frac{(ht)h(t-1)}{(2h)(h)}f_2$$

$$=\frac{(t-1)(t-2)}{2}f_0-\frac{(t)(t-2)}{1}f_1+\frac{(t)(t-1)}{2}f_2$$

Now,

$$\int_{x_0}^{x_2}f(x)dx=\int_{x_0}^{x_2}P_2(x)dx=\int_0^2 P_2(t)\quad hdt=h\int_0^2 P_2(t)hdt$$

$$=h\int_0^2\left(\frac{(t-1)(t-2)}{2}f_0-\frac{(t)(t-2)}{1}f_1+\frac{(t)(t-1)}{2}f_2\right)dx$$

$$=\frac{h}{2}\left[\int_0^2((t-1)(t-2))\ f_0\ dt-2\int_0^2((t)(t-2))f_1\ dt+\int_0^2((t)(t-1))f_2\ dt\right]$$

$$=\frac{h}{2}\left[\int_0^2(t^2-3t+2)\ f_0\ dt-2\int_0^2(t^2-2t)f_1\ dt+\int_0^2(t^2-t)\ f_2dt\right]$$

$$=\frac{h}{2}\left[\left|\frac{t^3}{3}-3\frac{t^2}{2}+2t\right|_0^2 f_0+\left|\frac{t^3}{3}-2\frac{t^2}{2}\right|_0^2 f_1+\left|\frac{t^3}{3}-\frac{t^2}{2}\right|_0^2 f_2\right]$$

$$=\frac{h}{2}\left[\frac{2}{3}f_0+\frac{8}{3}f_1+\frac{2}{3}f_2\right]$$

$$=\frac{h}{3}[f_0+4f_1+f_2]$$

By using Newton's forward difference formula

According to Newton's forward difference formula (7.2.2), we have:

$$P_2(p)=f_0+p\Delta f_0+\frac{p(p-1)}{2}\Delta^2 f_0$$

Now, $\int_{x_0}^{x_2}f(x)dx=\int_{x_0}^{x_2}P_2(x)dx$

$$= h\int_0^2 \left(f_0 + p\Delta f_0 + \frac{p(p-1)}{2}\Delta^2 f_0 \right) dp$$

$$= h\left(f_0 \left|p\right|_0^2 + \Delta f_0 \left|\frac{p^2}{2}\right|_0^2 + \Delta^2 f_0 \left|\frac{p^3}{6} - \frac{p^2}{4}\right|_0^2 \right)$$

$$= h\left(2f_0 + 2\Delta f_0 + \frac{1}{3}\Delta^2 f_0 \right)$$

$$= h\left(2f_0 + 2(f_1 - f_0) + \frac{1}{3}(f_2 - 2f_1 + f_0) \right)$$

$$= \frac{h}{3}(6f_0 + 6f_1 - 6f_0 + f_2 - 2f_1 + f_0) = \frac{h}{3}[f_o + 4\,f_1 + f_2] \qquad (7.7.1)$$

Equation (7.7.1) is Simpson's 1/3 rule to find the approximate value of integrand f(x) numerically.

7.7.1 Error Estimate in Simpson's 1/3 Rule

In Simpson's 1/3 rule, to find the integration of integrand f(x), we have approximated it by using interpolating function of degree two. We also considered that f(x) is sufficiently differentiable. Therefore, an error E_{SR} (f(x)) for this Newton-Cotes Quadrature formula has contained an approximate higher derivative. Therefore, by using equation (7.2.2), we have:

$$E_{SR}(f(x)) = h^3 f'''(c_n)\int_0^2 \frac{p(p-1)(p-2)}{3!}\,h\,dp = \frac{f''(c_n)}{6}\left|\frac{p^4}{4} - \frac{p^3}{1} + \frac{p^2}{1}\right|_0^2 = 0$$

where $a < c_n < b$

Since, third order error term is zero therefore we have to take the next higher term:

$$E_{SR}(f(x)) = h^4 f^{iv}(c_n)\int_0^2 \frac{p(p-1)(p-2)(p-3)}{4!}\,h\,dp$$

$$= h^5\frac{f''(c_n)}{24}\left|\frac{p^5}{5} - 6\frac{p^4}{4} + 11\frac{p^3}{3} - 6\frac{p^2}{2}\right|_0^2 = -\frac{h^5}{90}f^{iv}(c_n),$$

Thus, equation (7.7.1) can be written as:

$$\int_{x_0}^{x_2} f(x)\,dx = \frac{h}{3}(f_0 + 4f_1 + 4f_2) - \frac{h^5}{90} f^{iv}(c_n),$$

where,

$$E_{SR}(f(x)) = -\frac{h^5}{90} f^{iv}(c_n),$$

Example 7.6: Evaluate the integral $\int_0^{\Pi/2} (6+4 \cos x)dx$ by using Simpson's 1/3 rule.

Solution: According to Simpson's 1/3 rule, we have:

$$\int_{x_0}^{x_2} f(x)\,dx = \frac{h}{3}[f_0 + 4f_1 + f_2];$$

Now,

$$h = \frac{x_2 - x_0}{2} = \frac{\pi - 0}{2} = \frac{\pi}{2}$$

$$x_0 = 0,\ x_1 = \frac{\pi}{2},\ x_2 = \pi,$$

and f(x) = 6 + 4 cosx

$\Rightarrow f_0 = f(0) = 10,$

$$\Rightarrow f_1 = f\left(\frac{\pi}{2}\right) = 4\cos\frac{\pi}{2} + 6 = 6,$$

$\Rightarrow f_2 = f(\pi) = 4\cos\pi + 6 = 2;$

$$\therefore \int_{x_0}^{x_2} f(x)\,dx = \frac{h}{3}[f_0 + 4f_1 + f_2] = \frac{\pi}{6}[10 + 24 + 2] = 6\pi$$

Example 7.7: Use Gauss-Legendre's two point formula to find approximate value of $\int_{-1}^{1} \frac{1}{x+3}\,dx;$ dx; and compare the result with trapezoidal rule and with Simpson's 1/3 rule.

Solution: By using Gauss-Legendre's two point formula, we have:

$$\int_{-1}^{1} f(x)dx = f\left(\frac{1}{\sqrt{3}}\right) + f\left(-\frac{1}{\sqrt{3}}\right) \approx f(0.57735) + f(-0.57735);$$

Now, $f(x) = \dfrac{1}{x+3}$

$$\Rightarrow f(-0.57735) = \frac{1}{-0.57735+3} = 0.41277,$$

$$\Rightarrow f(0.57735) = \frac{1}{0.57735+3} = 0.27954,$$

$$\int_{-1}^{1} \frac{1}{x+3}dx = f(0.57735) + f(-0.57735) = 0.41277 + 0.27954 = 0.69231$$

Now, According to trapezoidal rule, we have:

$$\int_{x_0}^{x_1} f(x)dx = \frac{h}{2}[f_1 + f_0]$$

$x_0 = -1$, $x_1 = 1$, $h = x_1 - x_0 = 1 + 1 = 2$;

and $f(x) = \dfrac{1}{x+3}$

$$\Rightarrow f_0 = f(-1) = \frac{1}{-1+3} = 0.5,$$

$$\Rightarrow f_1 = f(1) = \frac{1}{1+3} = 0.25;$$

$$\therefore \int_{x_0}^{x_1} f(x)dx = \frac{h}{2}[f_1 + f_0] = \frac{2}{2}[0.5 + 0.25] = 0.75$$

Now, According to Simpson's 1/3 rule, we have:

$$\int_{x_0}^{x_2} f(x)dx = \frac{h}{3}[f_0 + 4f_1 + f_2];$$

$$h = \frac{x_2 - x_0}{2} = \frac{1+1}{2} = 1$$

$x_0 = -1, x_1 = 0, x_2 = 1,$

and $f(x) = \dfrac{1}{x+3}$

$$\Rightarrow f_0 = f(-1) = \frac{1}{-1+3} = 0.5,$$

$$\Rightarrow f_1 = f(0) = \frac{1}{0+3} = 0.33,$$

$$\Rightarrow f_2 = f(1) = \frac{1}{1+3} = 0.25;$$

$$\therefore \int_{x_0}^{x_2} f(x)dx = \frac{h}{3}[f_0 + 4f_1 + f_2] = \frac{1}{3}[0.5 + 4\times 0.33 + 0.25] = 0.69$$

Now, the exact value of the integral is:

$$\int_{-1}^{1} \frac{1}{x+3}\, dx = \left|\log|x+3|\right|_{-1}^{1} = \log 4 - \log 2 = 0.69315,$$

Evaluation of errors:

$e_{GL} = 0.69315 - 0.69231 = 0.00084,$

$e_{TR} = 0.69315 - 0.75 = -0.05685,$

$e_{SR} = 0.69315 - 0.69 = 0.00315,$

Hence, the comparison shows that Gauss-Legendre's two point formula gives better results than the others.

7.8 Simpson's 3/8 Rule

It is the Newton-Cote's closed Quadrature formula in which four nodes are taken i.e. $\{x_0, x_1, x_2, x_3\}$ and $n = 3$. Here, third order interpolation polynomial $P_3(x)$ is used to approximate the integrand f(x) over these four points.

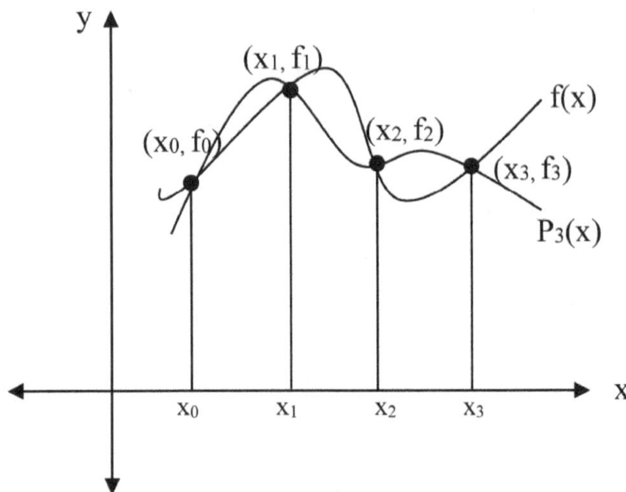

Fig. 7.4: Approximation of area by using three trapezoids

As shown in figure, f(x) is approximated by the combinations of trapezoids with horizontal base width h ($h = x_i - x_{i-1}$); i = 1, 2, 3 and sloped tops [$f_i = f(x_i)$, $f_{i-1} = f(x_{i-1})$]. Now, an interpolating polynomial throughout the points (x_0, $f(x_0)$), (x_1, $f(x_1)$), (x_2, $f(x_2)$) and (x_3, $f(x_3)$) is used for this approximation which can be defined in the following two ways:

By using Lagrange's Interpolation Formula:

According to Lagrange's interpolating polynomial (7.2.1) throughout these points is defined as follows:

$$P_3(x) = \frac{(x-x_3)(x-x_2)(x-x_1)}{(x_0-x_3)(x_0-x_2)(x_0-x_1)} f_0 + \frac{(x-x_3)(x-x_2)(x-x_0)}{(x_1-x_3)(x_1-x_2)(x_1-x_0)} f_1$$

$$+ \frac{(x-x_3)(x-x_1)(x-x_0)}{(x_2-x_3)(x_2-x_1)(x_2-x_0)} f_2 + \frac{(x-x_2)(x-x_1)(x-x_0)}{(x_3-x_2)(x_3-x_1)(x_3-x_0)} f_3$$

$$= \frac{(x-x_3)(x-x_2)(x-x_1)}{-6h^3} f_0 + \frac{(x-x_3)(x-x_2)(x-x_0)}{2h^3} f_1$$

$$+ \frac{(x-x_3)(x-x_1)(x-x_0)}{-2h^3} f_2 + \frac{(x-x_2)(x-x_1)(x-x_0)}{6h^3} f_3$$

Now,

$$\int_{x_0}^{x_3} f(x)\,dx = \int_{x_0}^{x_3} P_3(x)\,dx$$

$$= \int_{x_0}^{x_3} \left(\begin{array}{c} \dfrac{(x-x_3)(x-x_2)(x-x_1)}{-6h^3} f_0 + \dfrac{(x-x_3)(x-x_2)(x-x_0)}{2h^3} f_1 + \\ \dfrac{(x-x_3)(x-x_1)(x-x_0)}{-2h^3} f_2 + \dfrac{(x-x_2)(x-x_1)(x-x_0)}{6h^3} f_3 \end{array} \right) dx$$

After simplification; we get:

$$\int_{x_0}^{x_3} f(x)\,dx = \frac{3h}{8}\left[f_0 + 3f_1 + 3f_2 + f_3\right]$$

Alternate method of Simplification

Similarly, for integration nodes $\{x_0, x_1, x_2, x_3\}$; we can use the linear transformation as: $x = x_0 + ht \Rightarrow dx = dt$ and $t \to 0$ as $x \to x_0$; $t \to 3$ as $x \to x_3$

also, $x - x_0 = ht$; $x - x_1 = h(t-1)$; $x - x_2 = h(t-2)$; $x - x_3 = h(t-3)$;

$(x_3 - x_0) = 3h$; $(x_3 - x_1) = 2h$;$(x_3 - x_2) = 1h$; $(x_2 - x_0) = 2h$;
$(x_2 - x_1) = h$; $(x_1 - x_0) = h$

$$\therefore P_3(x) = \frac{(x-x_3)(x-x_2)(x-x_1)}{(x_0-x_3)(x_0-x_2)(x_0-x_1)} f_0 + \frac{(x-x_3)(x-x_2)(x-x_0)}{(x_1-x_3)(x_1-x_2)(x_1-x_0)} f_1$$

$$+ \frac{(x-x_3)(x-x_1)(x-x_0)}{(x_2-x_3)(x_2-x_1)(x_2-x_0)} f_2 + \frac{(x-x_2)(x-x_1)(x-x_0)}{(x_3-x_2)(x_3-x_1)(x_3-x_0)} f_3$$

becomes as follows:

$$P_3(x) = \frac{h(t-3)h(t-2)h(t-1)}{-6h^3} f_0 + \frac{h(t-3)h(t-2)ht}{2h^3} f_1$$

$$+ \frac{h(t-3)h(t-1)ht}{-2h^3} f_2 + \frac{h(t-2)h(t-1)ht}{6h^3} f_3$$

$$= \frac{(t-3)(t-2)(t-1)}{-6} f_0 + \frac{(t-3)(t-2)t}{2} f_1 + \frac{(t-3)(t-1)t}{-2} f_2 + \frac{(t-2)(t-1)t}{6} f_3$$

Now,

$$\int_{x_0}^{x_3} f(x)dx = \int_{x_0}^{x_3} P_3(x)dx = \int_0^3 P_3(t) \quad hdt = h\int_0^3 P_3(t)hdt$$

$$= h\int_0^3 \left(\frac{(t-3)(t-2)(t-1)}{-6} f_0 + \frac{(t-3)(t-2)t}{2} f_1 + \frac{(t-3)(t-1)t}{-2} f_2 + \frac{(t-2)(t-1)t}{6} f_3 \right) dx$$

$$= \frac{h}{6}\left[-\int_0^3 ((t-3)(t-2)(t-1))\ f_0\ dt + 3\int_0^3 ((t-3)(t-2)t)\ f_1 dt - 3\int_0^3 ((t-3)(t-1)t)\ f_2 dt + \int_0^3 ((t-2)(t-1)t) f_3\ dt \right]$$

$$= \frac{h}{6}\left[-\int_0^3 (t^3 - 6t^2 + 11t - 6) f_0\ dt + 3\int_0^3 (t^3 - 5t^2 + 6t) f_1\ dt - 3\int_0^3 (t^3 - 4t^2 + 3t) f_2\ dt + \int_0^3 (t^3 - 3t^2 + 2t) f_3 dt \right]$$

$$= \frac{h}{6}\left[-\left| \frac{t^4}{4} - 6\frac{t^3}{3} + 11\frac{t^2}{2} - 6t \right|_0^3 f_0 + 3\left| \frac{t^4}{4} - 5\frac{t^3}{3} + 6\frac{t^2}{2} \right|_0^3 f_1 - 3\left| \frac{t^4}{4} - 4\frac{t^3}{3} + 3\frac{t^2}{2} \right|_0^3 f_2 + \left| \frac{t^4}{4} - 3\frac{t^3}{3} + 2\frac{t^2}{2} \right|_0^3 f_3 \right]$$

$$= \frac{h}{6}\left[\frac{18}{8} f_0 + \frac{54}{8} f_1 + \frac{54}{8} f_2 + \frac{18}{8} f_3 \right]$$

$$= \frac{3h}{8}[f_0 + 3f_1 + 3f_2 + f_3]$$

By using Newton's forward difference formula:

According to Newton's forward difference formula (7.2.2), we have:

$$P_3(p) = f_0 + p\Delta f_0 + \frac{p(p-1)}{2!}\Delta^2 f_0 + \frac{p(p-1)(p-2)}{3!}\Delta^3 f_0$$

Now, $\int_{x_0}^{x_3} f(x)dx = \int_{x_0}^{x_3} P_3(x)dx$

$$= h\int_0^3 \left(f_0 + p\Delta f_0 + \frac{p(p-1)}{2!}\Delta^2 f_0 + + \frac{p(p-1)(p-2)}{3!}\Delta^3 f_0 \right) dp$$

$$= h\left(f_0 |p|_0^3 + \Delta f_0 \left| \frac{p^2}{2} \right|_0^3 + \Delta^2 f_0 \left| \frac{p^3}{6} - \frac{p^2}{4} \right|_0^3 + \frac{\Delta^3 f_0}{6} \left| \frac{p^4}{4} - \frac{3p^3}{3} + \frac{2p^2}{2} \right|_0^3 \right)$$

$$= h\left(f_0 \left| p \right|_0^3 + \Delta f_0 \left| \frac{p^2}{2} \right|_0^3 + \Delta^2 f_0 \left| \frac{p^3}{6} - \frac{p^2}{4} \right|_0^3 + \frac{\Delta^3 f_0}{6} \left| \frac{p^4}{4} - \frac{3p^3}{3} + \frac{2p^2}{2} \right|_0^3 \right)$$

$$= h\left(3f_0 + \frac{9}{2}(f_1 - f_0) + \frac{9}{4}(f_2 - 2f_1 + f_0) + \frac{9}{24}(f_3 - 3f_2 + 3f_1 - f_0) \right)$$

$$= \frac{9h}{24}\left(8f_0 + 12(f_1 - f_0) + 6(f_2 - 2f_1 + f_0) + 1(f_3 - 3f_2 + 3f_1 - f_0)\right)$$

$$= \frac{3h}{8}\left[f_0 + 3f_1 + 3f_2 + f_3\right] \tag{7.8.1}$$

Equation (7.8.1) is Simpson's 3/8 rule to find the approximate value of integrand f(x) numerically.

Example 7.8: Evaluate the integral $\int_0^1 \frac{1}{1+x^2}\,dx$ by using Simpson's 3/8 rule.

Solution: According to Simpson's 3/8 rule, we have:

$$\int_{x_0}^{x_3} f(x)\,dx = \frac{3h}{8}\left[f_0 + 3f_1 + 3f_2 + f_3\right];$$

Now,

$$h = \frac{x_3 - x_0}{3} = \frac{1-0}{3} = \frac{1}{3}$$

$$x_0 = 0,\ x_1 = \frac{1}{3},\ x_2 = \frac{2}{3},\ x_3 = 1,$$

and $f(x) = \frac{1}{1+x^2}$

$$\Rightarrow f_0 = f(0) = \frac{1}{1+0} = 1,$$

$$\Rightarrow f_1 = f\left(\frac{1}{3}\right) = \frac{1}{1+\frac{1}{9}} = \frac{9}{10} = 0.9,$$

$$\Rightarrow f_2 = f\left(\frac{2}{3}\right) = \frac{1}{1+\frac{4}{9}} = \frac{9}{13} = 0.69231,$$

$$\Rightarrow f_3 = f(1) = \frac{1}{1+1} = \frac{1}{2} = 0.5;$$

$$\therefore \int_{x_0}^{x_3} f(x)dx = \frac{3h}{8}[f_0 + 3f_1 + 3f_2 + f_3] = \frac{3}{8} \times \frac{1}{3}[1 + 2.7 + 2.07693 + 0.5] \approx 0.78462$$

7.9 Boole's and Weddle's Rules

George Boole introduced a numerical method of integration which is called as Boole's rule. This method is a Newton-Cote's Quadrature formula with n = 4. Thomas Weddle also presented Newton-Cote's Quadrature formula with n = 6 which is known as Weddle's rule. These rules are explained in the following way:

7.9.1 Boole's Rule

It is the Newton-Cote's closed Quadrature formula in which five nodes are taken i.e. $\{x_0, x_1, x_2, x_3, x_4\}$ and n = 4. Here, fourth order interpolation polynomial P_4 (x) is used to approximate the integrand f(x) over five points.

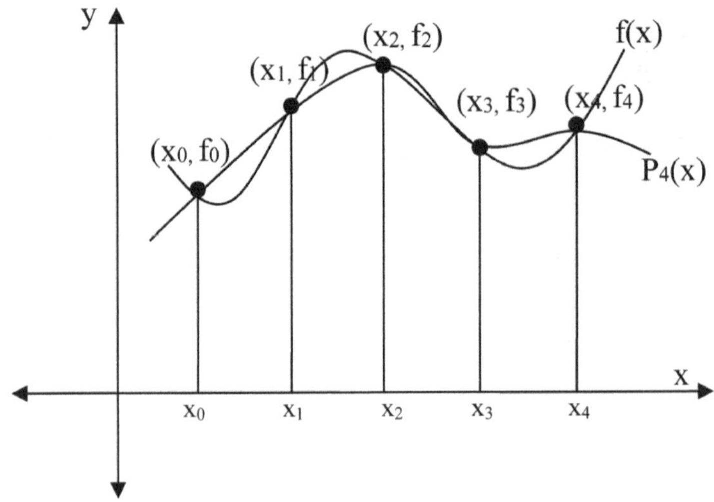

As shown in figure, f(x) is approximated by the combinations of trapezoids with horizontal base width h ($h = x_i - x_{i-1}$); i = 1, 2, 3, 4 and sloped tops [$f_i = f(x_i)$, $f_{i-1}) = f(x_{i-1})$]. Now, an interpolating polynomial throughout the points $(x_0, f(x_0))$, $(x_1, f(x_1))$, $(x_2, f(x_2))$, $(x_3, f(x_3))$ and $(x_4, f(x_4))$ is used for this approximation which can be defined in the following two ways:

By using Lagrange's Interpolation Formula:

According to Lagrange's interpolating polynomial (7.2.1) throughout these points is defined as follows:

$$P_4(x) = \frac{(x-x_4)(x-x_3)(x-x_2)(x-x_1)}{(x_0-x_4)(x_0-x_3)(x_0-x_2)(x_0-x_1)} f_0$$
$$+ \frac{(x-x_4)(x-x_3)(x-x_2)(x-x_0)}{(x_1-x_4)(x_1-x_3)(x_1-x_2)(x_1-x_0)} f_1$$
$$+ \frac{(x-x_4)(x-x_3)(x-x_1)(x-x_0)}{(x_2-x_4)(x_2-x_3)(x_2-x_1)(x_2-x_0)} f_2$$
$$+ \frac{(x-x_4)(x-x_2)(x-x_1)(x-x_0)}{(x_3-x_4)(x_3-x_2)(x_3-x_1)(x_3-x_0)} f_3$$
$$+ \frac{(x-x_3)(x-x_2)(x-x_1)(x-x_0)}{(x_4-x_3)(x_4-x_2)(x_4-x_1)(x_4-x_0)} f_4$$

$$= \frac{(x-x_4)(x-x_3)(x-x_2)(x-x_1)}{24h^4} f_0 + \frac{(x-x_4)(x-x_3)(x-x_2)(x-x_0)}{-6h^4} f_1$$
$$+ \frac{(x-x_4)(x-x_3)(x-x_1)(x-x_0)}{4h^4} f_2 + \frac{(x-x_4)(x-x_2)(x-x_1)(x-x_0)}{-6h^4} f_3$$
$$+ \frac{(x-x_3)(x-x_2)(x-x_1)(x-x_0)}{24h^4} f_4$$

Now,

$$\int_{x_0}^{x_4} f(x)\,dx = \int_{x_0}^{x_4} P_4(x)\,dx$$

$$= \int_{x_0}^{x_4} \left(\frac{(x-x_4)(x-x_3)(x-x_2)(x-x_1)}{24h^4} f_0 \right.$$

$$+ \frac{(x-x_4)(x-x_3)(x-x_2)(x-x_0)}{-6h^4} f_1$$

$$+ \frac{(x-x_4)(x-x_3)(x-x_1)(x-x_0)}{4h^4} f_2$$

$$+ \frac{(x-x_4)(x-x_2)(x-x_1)(x-x_0)}{-6h^4} f_3$$

$$\left. + \frac{(x-x_3)(x-x_2)(x-x_1)(x-x_0)}{24h^4} f_4 \right) dx$$

After simplification; we get:

$$\int_{x_0}^{x_4} f(x)\,dx = \frac{2h}{45}\left[7f_0 + 32f_1 + 12f_2 + 32f_3 + 7f_4\right]$$

Alternate method of Simplification

Similarly, for integration nodes $\{x_0, x_1, x_2, x_3, x_4\}$; we can use the linear transformation as: $x = x_0 + ht \to dx = dt$ and $t \to 0$ as $x \to x_0$; $t \to 4$ as $x \to x_4$

also,

$x - x_0 = ht$; $x - x_1 = h(t-1)$; $x - x_2 = h(t-2)$; $x - x_3 = h(t-3)$; $x - x_4 = h(t-4)$;

and $(x_4 - x_0) = 4h$; $(x_4 - x_1) = 3h$; $(x_4 - x_2) = 2h$; $(x_4 - x_3) = 1h$;

$(x_3 - x_0) = 3h$; $(x_3 - x_1) = 2h$; $(x_3 - x_2) = 1h$;

$(x_2 - x_0) = 2h$; $(x_2 - x_1) = h$; $(x_1 - x_0) = h$

$$P_4(x) = \left(\frac{(x-x_4)(x-x_3)(x-x_2)(x-x_1)}{24h^4} f_0 + \frac{(x-x_4)(x-x_3)(x-x_2)(x-x_0)}{-6h^4} f_1 \right.$$

$$+ \frac{(x-x_4)(x-x_3)(x-x_1)(x-x_0)}{4h^4} f_2 + \frac{(x-x_4)(x-x_2)(x-x_1)(x-x_0)}{-6h^4} f_3$$

$$\left. + \frac{(x-x_3)(x-x_2)(x-x_1)(x-x_0)}{24h^4} f_4 \right)$$

and it becomes as follows:

$$\therefore P_4(x) = \frac{h(t-4)h(t-3)h(t-2)h(t-1)}{24h^4} f_0 + \frac{h(t-4)h(t-3)h(t-2)ht}{-6h^4} f_1$$
$$+\frac{h(t-4)h(t-3)h(t-1)ht}{4h^4} f_2 + \frac{h(t-4)h(t-2)h(t-1)ht}{-6h^4} f_3$$
$$+\frac{h(t-3)h(t-2)h(t-1)ht}{24h^4} f_4$$

$$=\frac{(t-4)(t-3)(t-2)(t-1)}{24} f_0 + \frac{(t-4)(t-3)(t-2)t}{-6} f_1 + \frac{(t-4)(t-3)(t-1)t}{4} f_2$$
$$+\frac{(t-4)(t-2)(t-1)t}{-6} f_3 + \frac{(t-3)(t-2)(t-1)t}{24} f_4$$

Now,

$$\int_{x_0}^{x_4} f(x)\,dx = \int_{x_0}^{x_4} P_4(x)\,dx = \int_0^4 P_4(t) \quad hdt = h\int_0^4 P_4(t) \quad h\,dt$$

$$= h\int_0^4 \left(\frac{(t-4)(t-3)(t-2)(t-1)}{24} f_0 + \frac{(t-4)(t-3)(t-2)t}{-6} f_1 \right.$$
$$\left. + \frac{(t-4)(t-3)(t-1)t}{4} f_2 + \frac{(t-4)(t-2)(t-1)t}{-6} f_3 + \frac{(t-3)(t-2)(t-1)t}{24} f_4 \right) dx$$

$$= \frac{h}{24}\left[\int_0^4 ((t-4)(t-3)(t-2)(t-1))\ f_0\ dt - 4\int_0^4 ((t-4)(t-3)(t-2)t)\ f_1 dt \right.$$
$$+6\int_0^4 ((t-4)(t-3)(t-1)t)\ f_2 dt - 4\int_0^4 ((t-4)(t-2)(t-1)t) f_3\ dt$$
$$\left. + \int_0^4 ((t-3)(t-2)(t-1)t) f_4\ dt \right]$$

$$= \frac{h}{24}\left[\int_0^4 \left(t^4 - 10t^3 + 35t^2 - 50t + 24\right) f_0\, dt - 4\int_0^4 \left(t^4 - 9t^3 + 26t^2 - 24t\right) f_1 dt\right.$$

$$+6\int_0^4 \left(t^4 - 8t^3 + 19t^2 - 12t\right) f_2 dt - 4\int_0^4 \left(t^4 - 7t^3 + 14t^2 - 8t\right) f_3\, dt$$

$$\left.+\int_0^4 \left(t^4 - 6t^3 + 11t^2 - 6t\right) f_4\, dt\right]$$

$$= \frac{h}{24}\left[\left|\frac{t^5}{5} - 10\frac{t^4}{4} + 35\frac{t^3}{3} - 50\frac{t^2}{2} + 24t\right|_0^4 f_0 - 4\left|\frac{t^5}{5} - 9\frac{t^4}{4} + 26\frac{t^3}{3} - 24\frac{t^2}{2}\right|_0^4 f_1\right.$$

$$+6\left|\frac{t^5}{5} - 8\frac{t^4}{4} + 19\frac{t^3}{3} - 12\frac{t^2}{2}\right|_0^4 f_2 - 4\left|\frac{t^5}{5} - 7\frac{t^4}{4} + 14\frac{t^3}{3} - 8\frac{t^2}{2}\right|_0^4 f_3$$

$$\left.+\left|\frac{t^5}{5} - 6\frac{t^4}{4} + 11\frac{t^3}{3} - 6\frac{t^2}{2}\right|_0^4 f_4\right]$$

$$= \frac{h}{24}\left[\frac{336}{45} f_0 + \frac{1536}{45} f_1 + \frac{576}{45} f_2 + \frac{1536}{45} f_3 + \frac{336}{45} f_4\right]$$

$$= \frac{2h}{45}\left[7f_0 + 32f_1 + 12f_2 + 32f_3 + 7f_4\right]$$

By using Newton's forward difference formula:

According to Newton's forward difference formula (7.2.2), we have:

$$P_4(p) = f_0 + p\Delta f_0 + \frac{p(p-1)}{2!}\Delta^2 f_0 + \frac{p(p-1)(p-2)}{3!}\Delta^3 f_0$$

$$+\frac{p(p-1)(p-2)(p-3)}{4!}\Delta^4 f_0$$

Now, $\int_{x_0}^{x_4} f(x)dx = \int_{x_0}^{x_4} P_4(x)dx$

$$= h\int_0^4 \left(f_0 + p\Delta f_0 + \frac{p(p-1)}{2!}\Delta^2 f_0 + + \frac{p(p-1)(p-2)}{3!}\Delta^3 f_0 \right.$$

$$\left. + \frac{p(p-1)(p-2)(p-3)}{4!}\Delta^4 f_0 \right) dp$$

$$= h\left(f_0 \left|p\right|_0^4 + \Delta f_0 \left|\frac{p^2}{2}\right|_0^4 + \frac{\Delta^2 f_0}{2}\left|\frac{p^3}{3} - \frac{p^2}{2}\right|_0^4 + \frac{\Delta^3 f_0}{6}\left|\frac{p^4}{4} - \frac{3p^3}{3} + \frac{2p^2}{2}\right|_0^4 \right.$$

$$\left. + \frac{\Delta^4 f_0}{24}\left|\frac{p^5}{5} - 6\frac{p^4}{4} + 11\frac{p^3}{3} - 6\frac{p^2}{2}\right|_0^4 \right)$$

$$= h\left(4f_0 + 8\Delta f_0 + \frac{20}{3}\Delta^2 f_0 + \frac{8}{3}\Delta^3 f_0 + \frac{14}{45}\Delta^4 f_0 \right)$$

$$= h\left(4f_0 + 8(f_1 - f_0) + \frac{20}{3}(f_2 - 2f_1 + f_0) + \frac{8}{3}(f_3 - 3f_2 + 3f_1 - f_0) + \right.$$

$$\left. \frac{14}{45}(f_4 - 4f_3 + 6f_2 - 4f_1 + f_0) \right)$$

$$= \frac{2h}{45}\left(90f_0 + 180(f_1 - f_0) + 150(f_2 - 2f_1 + f_0) + 60(f_3 - 3f_2 + 3f_1 - f_0)\right.$$

$$\left. +7(f_4 - 4f_3 + 6f_2 - 4f_1 + f_0)\right)$$

$$= \frac{2h}{45}[7f_0 + 32f_1 + 12f_2 + 32f_3 + 7f_4] \qquad (7.9.1.1)$$

Equation (7.9.1.1) is Boole's rule to find the approximate value of integrand f(x) numerically.

Example 7.9: Evaluate the integral $\int_0^2 e^{x^2} - 1\, dx$ by using Boole's rule.

Solution: According to Boole's rule, we have:

$$\int_{x_0}^{x_4} f(x)dx = \frac{2h}{45}[7f_0 + 32f_1 + 12f_2 + 32f_3 + 7f_4];$$

Now,

$$h = \frac{x_4 - x_0}{4} = \frac{2-0}{4} = 0.5;$$

$x_0 = 0, \quad x_1 = 0.5, \quad x_2 = 1.0, \quad x_3 = 1.5, \quad x_4 = 2.0$

and f(x) = $e^{x^2} - 1$;

$\Rightarrow$ f0 = f(0) = e^0 – 1 = 1–1 = 0,

$\Rightarrow f_1$ = f(0.5) = $e^{0.25}$ – 1 = – 0.32043,

$\Rightarrow f_2$ = f(1) = e^1 – 1 = 1.71828,

$\Rightarrow f_3$ = f(1.5) = $e^{2.25}$ – 1 = 5.11613,

$\Rightarrow f_4$ = f(2) = e^4 – 1 = 9.87313;

$$\therefore \int_{x_0}^{x_4} f(x)dx = \frac{2h}{45}[7f_0 + 32f_1 + 12f_2 + 32f_3 + 7f_4]$$

$= \frac{2}{45} \times \frac{1}{2}$ [7 × 0 + 32 × (-0.32043) + 12 × 1.71828 + 32 × 5.11613 + 7 × 9.87313] ≈ 5.4043

7.9.2 Weddle's Rule

It is the Newton-Cote's closed Quadrature formula in which seven nodes are taken i.e. $\{x_0, x_1, x_2, x_3, x_4, x_5, x_6\}$ and n=6. Here, sixth order interpolation polynomial P_6 (x) is used to approximate the integrand f(x) over these seven points.

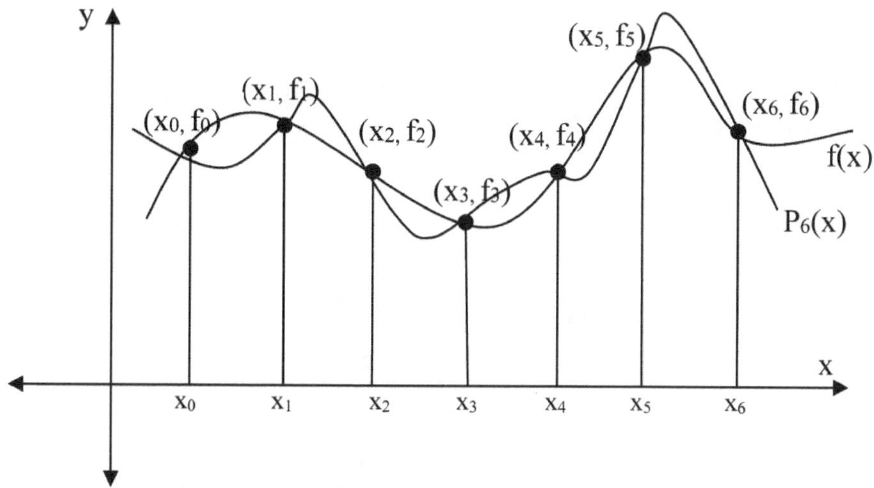

Fig. 7.6: Approximation of area by using six trapezoids

As shown in figure, f(x) is approximated by the combinations of trapezoids with horizontal base width h ($h = x_i - x_{i-1}$); i = 1,2,3,4,5,6 and sloped tops [$f_i = f(x_i)$, $f_{i-1} = f(x_{i-1})$]. Now, an interpolating polynomial throughout the points $(x_0, f(x_0))$, $(x_1, f(x_1))$, $(x_2, f(x_2))$, $(x_3, f(x_3))$, $(x_4, f(x_4))$, $(x_5, f(x_5))$ and $(x_6, f(x_6))$ is used for this approximation which can be defined in the following two ways:

By using Lagrange's Interpolation Formula:

According to Lagrange's interpolating polynomial (7.2.1) throughout the points is defined as follows:

$$P_6(x)=\frac{(x-x_6)(x-x_5)(x-x_4)(x-x_3)(x-x_2)(x-x_1)}{(x_0-x_6)(x_0-x_5)(x_0-x_4)(x_0-x_3)(x_0-x_2)(x_0-x_1)}f_0$$

$$+\frac{(x-x_6)(x-x_5)(x-x_4)(x-x_3)(x-x_2)(x-x_0)}{(x_1-x_6)(x_1-x_5)(x_1-x_4)(x_1-x_3)(x_1-x_2)(x_1-x_0)}f_1$$

$$+\frac{(x-x_6)(x-x_5)(x-x_4)(x-x_3)(x-x_1)(x-x_0)}{(x_2-x_6)(x_2-x_5)(x_2-x_4)(x_2-x_3)(x_2-x_1)(x_2-x_0)}f_2$$

$$+\frac{(x-x_6)(x-x_5)(x-x_4)(x-x_2)(x-x_1)(x-x_0)}{(x_3-x_6)(x_3-x_5)(x_3-x_4)(x_3-x_2)(x_3-x_1)(x_3-x_0)}f_3$$

$$+\frac{(x-x_6)(x-x_5)(x-x_3)(x-x_2)(x-x_1)(x-x_0)}{(x_4-x_6)(x_4-x_5)(x_4-x_3)(x_4-x_2)(x_4-x_1)(x_4-x_0)}f_4$$

$$+\frac{(x-x_6)(x-x_4)(x-x_3)(x-x_2)(x-x_1)(x-x_0)}{(x_5-x_6)(x_5-x_4)(x_5-x_3)(x_5-x_2)(x_5-x_1)(x_5-x_0)}f_5$$

$$+\frac{(x-x_5)(x-x_4)(x-x_3)(x-x_2)(x-x_1)(x-x_0)}{(x_6-x_5)(x_6-x_4)(x_6-x_3)(x_6-x_2)(x_6-x_1)(x_6-x_0)}f_6$$

$$=\frac{(x-x_6)(x-x_5)(x-x_4)(x-x_3)(x-x_2)(x-x_1)}{720h^6}f_0$$

$$+\frac{(x-x_6)(x-x_5)(x-x_4)(x-x_3)(x-x_2)(x-x_0)}{-120h^6}f_1$$

$$+\frac{(x-x_6)(x-x_5)(x-x_4)(x-x_3)(x-x_1)(x-x_0)}{48h^6}f_2$$

$$+\frac{(x-x_6)(x-x_5)(x-x_4)(x-x_2)(x-x_1)(x-x_0)}{-36h^6}f_3$$

$$+\frac{(x-x_6)(x-x_5)(x-x_3)(x-x_2)(x-x_1)(x-x_0)}{48h^6}f_4$$

$$+\frac{(x-x_6)(x-x_4)(x-x_3)(x-x_2)(x-x_1)(x-x_0)}{-120h^6}f_5$$

$$+\frac{(x-x_5)(x-x_4)(x-x_3)(x-x_2)(x-x_1)(x-x_0)}{720h^6}f_6$$

Now,

$$\int_{x_0}^{x_6} f(x)dx = \int_{x_0}^{x_6} P_6(x)dx$$

$$= \int \left(\frac{(x-x_6)(x-x_5)(x-x_4)(x-x_3)(x-x_2)(x-x_1)}{720h^6} f_0 \right.$$
$$+ \frac{(x-x_6)(x-x_5)(x-x_4)(x-x_3)(x-x_2)(x-x_0)}{-120h^6} f_1$$
$$+ \frac{(x-x_6)(x-x_5)(x-x_4)(x-x_3)(x-x_1)(x-x_0)}{48h^6} f_2$$
$$+ \frac{(x-x_6)(x-x_5)(x-x_4)(x-x_2)(x-x_1)(x-x_0)}{-36h^6} f_3$$
$$+ \frac{(x-x_6)(x-x_5)(x-x_3)(x-x_2)(x-x_1)(x-x_0)}{48h^6} f_4$$
$$+ \frac{(x-x_6)(x-x_4)(x-x_3)(x-x_2)(x-x_1)(x-x_0)}{-120h^6} f_5$$
$$\left. + \frac{(x-x_5)(x-x_4)(x-x_3)(x-x_2)(x-x_1)(x-x_0)}{720h^6} f_6 \right) dx$$

After simplification; we get:

$$\int_{x_0}^{x_6} f(x)dx = \frac{3h}{10}[f_0 + 5f_1 + f_2 + 6f_3 + f_4 + 5f_5 + f_6] \qquad (7.9.2.1)$$

Alternate method of Simplification

Similarly, for integration nodes $\{x_0, x_1, x_2, x_3, x_4, x_5, x_6\}$; we can use the linear transformation as: $x = x_0 + ht \Rightarrow dx = dt$ and $t \to 0$ as $x \to x_0$; $t \to 6$ as $x \to x_6$

also,

$x - x_0 = ht$, $x - x_1 = h(t-1)$, $x - x_2 = h(t-2)$, $x - x_3 = h(t-3)$;

$x - x_4 = h(t-4)$, $x - x_5 = h(t-5)$, $x - x_6 = h(t-6)$;

and $(x_6 - x_0) = 6h$, $(x_6 - x1) = 5h$, $(x_6 - x_2) = 4h$, $(x_6 - x_3) = 3h$, $(x_6 - x_4) = 2h$; $(x_6 - x_5) = 1h$;

$(x_5 - x_0) = 5h$,$(x_5 - x_1) = 4h$, $(x_5 - x_2) = 3h$, $(x_5 - x_3) = 2h$,$(x_5 - x_4) = 1h$;

$(x_4 - x_0) = 4h$, $(x_4 - x_1) = 3h$, $(x_4 - x_2) = 2h$, $(x_4 - x_3) = 1h$;

$(x_3 - x_0) = 3h$, $(x_3 - x_1) = 2h$, $(x^3 - x_2) = 1h$;

$(x_2 - x_0) = 2h$, $(x_2 - x_1) = h$, $(x_1 - x_0) = h$;

$$P_6(x) = \left(\frac{(x-x_6)(x-x_5)(x-x_4)(x-x_3)(x-x_2)(x-x_1)}{720h^6} f_0 \right.$$

$$+ \frac{(x-x_6)(x-x_5)(x-x_4)(x-x_3)(x-x_2)(x-x_0)}{-120h^6} f_1$$

$$+ \frac{(x-x_6)(x-x_5)(x-x_4)(x-x_3)(x-x_1)(x-x_0)}{48h^6} f_2$$

$$+ \frac{(x-x_6)(x-x_5)(x-x_4)(x-x_2)(x-x_1)(x-x_0)}{-36h^6} f_3$$

$$+ \frac{(x-x_6)(x-x_5)(x-x_3)(x-x_2)(x-x_1)(x-x_0)}{48h^6} f_4$$

$$+ \frac{(x-x_6)(x-x_4)(x-x_3)(x-x_2)(x-x_1)(x-x_0)}{-120h^6} f_5$$

$$\left. + \frac{(x-x_5)(x-x_4)(x-x_3)(x-x_2)(x-x_1)(x-x_0)}{720h^6} f_6 \right)$$

and it becomes as follows:

$$\therefore P_6(x) = \frac{h(t-6)h(t-5)h(t-4)h(t-3)h(t-2)h(t-1)}{720h^6} f_0$$
$$+\frac{h(t-6)h(t-5)h(t-4)h(t-3)h(t-2)h(t)}{-120h^6} f_1$$
$$+\frac{h(t-6)h(t-5)h(t-4)h(t-3)h(t-1)h(t)}{48h^6} f_2$$
$$+\frac{h(t-6)h(t-5)h(t-4)h(t-2)h(t-1)h(t)}{-36h^6} f_3$$
$$+\frac{h(t-6)h(t-5)h(t-3)h(t-2)h(t-1)h(t)}{48h^6} f_4$$
$$+\frac{h(t-6)h(t-4)h(t-3)h(t-2)h(t-1)h(t)}{-120h^6} f_5$$
$$+\frac{h(t-5)h(t-4)h(t-3)h(t-2)h(t-1)h(t)}{720h^6} f_6$$

$$= \frac{(t-6)(t-5)(t-4)(t-3)(t-2)(t-1)}{720} f_0 + \frac{(t-6)(t-5)(t-4)(t-3)(t-2)(t)}{-120} f_1$$
$$+\frac{(t-6)(t-5)(t-4)(t-3)(t-1)(t)}{48} f_2 + \frac{(t-6)(t-5)(t-4)(t-2)(t-1)(t)}{-36} f_3$$
$$+\frac{(t-6)(t-5)(t-3)(t-2)(t-1)(t)}{48} f_4 + \frac{(t-6)(t-4)(t-3)(t-2)(t-1)(t)}{-120} f_5$$
$$+\frac{(t-5)(t-4)(t-3)(t-2)(t-1)(t)}{720} f_6$$

Now,

$$\int_{x_0}^{x_6} f(x)\,dx = \int_{x_0}^{x_6} P_6(x)\,dx = \int_0^6 P_6(t) \quad hdt = h\int_0^6 P_6(t) \quad h\,dt$$

$$=h\int_0^6\left(\frac{(t-6)(t-5)(t-4)(t-3)(t-2)(t-1)}{720}f_0+\right.$$

$$\frac{(t-6)(t-5)(t-4)(t-3)(t-2)(t)}{-120}f_1+\frac{(t-6)(t-5)(t-4)(t-3)(t-1)(t)}{48}f_2$$

$$+\frac{(t-6)(t-5)(t-4)(t-2)(t-1)(t)}{-36}f_3+\frac{(t-6)(t-5)(t-3)(t-2)(t-1)(t)}{48}f_4$$

$$\left.+\frac{(t-6)(t-4)(t-3)(t-2)(t-1)(t)}{-120}f_5+\frac{(t-5)(t-4)(t-3)(t-2)(t-1)(t)}{720}f_6\right)dx$$

$$=\frac{h}{720}\left[\int_0^6((t-6)(t-5)(t-4)(t-3)(t-2)(t-1))\ f_0\ dt\right.$$

$$-6\int_0^6((t-6)(t-5)(t-4)(t-3)(t-2)(t))\ f_1dt$$

$$+15\int_0^6((t-6)(t-5)(t-4)(t-3)(t-1)(t))\ f_2dt$$

$$-20\int_0^6((t-6)(t-5)(t-4)(t-2)(t-1)(t))f_3\ dt$$

$$+15\int_0^6((t-6)(t-5)(t-3)(t-2)(t-1)(t))f_4\ dt$$

$$+6\int_0^6((t-6)(t-4)(t-3)(t-2)(t-1)(t))f_5\ dt$$

$$\left.+\int_0^6((t-5)(t-4)(t-3)(t-2)(t-1)(t))f_6\ dt\right]$$

$$=\frac{h}{720}\left[\int_0^6((t-6)(t-5)(t-4)(t-3)(t-2)(t-1))\ f_0\ dt\right.$$

$$-6\int_0^6((t-6)(t-5)(t-4)(t-3)(t-2)(t))\ f_1dt$$

$$+15\int_0^6((t-6)(t-5)(t-4)(t-3)(t-1)(t))\ f_2dt$$

$$-20\int_0^6((t-6)(t-5)(t-4)(t-2)(t-1)(t))f_3\ dt$$

$$+15\int_0^6 ((t-6)(t-5)(t-3)(t-2)(t-1)(t))f_4\ dt$$

$$+6\int_0^6 ((t-6)(t-4)(t-3)(t-2)(t-1)(t))f_5\ dt$$

$$\left.+\int_0^6 ((t-5)(t-4)(t-3)(t-2)(t-1)(t))f_6\ dt\right]$$

$$=\frac{h}{720}\left[\int_0^6 \left(t^6-21t^5+175t^4-735t^3+1624t^2-1764t+720\right)\ f_0\ dt\right.$$

$$-6\int_0^6 \left(t^6-20t^5+155t^4-580t^3+1044t^2-720t\right)\ f_1 dt$$

$$+15\int_0^6 \left(t^6-19t^5+137t^4-461t^3+702t^2-360t\right)\ f_2 dt$$

$$-20\int_0^6 \left(t^6-18t^5+121t^4-372t^3+508t^2-240t\right)f_3\ dt$$

$$+15\int_0^6 \left(t^6-17t^5+107t^4-307t^3+396t^2-180t\right)f_4\ dt$$

$$+6\int_0^6 \left(t^6-16t^5+95t^4-260t^3+324t^2-144t\right)f_5\ dt$$

$$\left.+\int_0^6 \left(t^6-15t^5+85t^4-225t^3+274t^2-120t\right)f_6\ dt\right]$$

$$\frac{h}{720}\Bigg(\left|\left(\frac{t^7}{7}-21\frac{t^6}{6}+175\frac{t^5}{5}-735\frac{t^4}{4}+1624\frac{t^3}{3}-1764\frac{t^2}{2}+720t\right)f_0\right|_0^6$$

$$-6\left|\left(\frac{t^7}{7}-20\frac{t^6}{6}+155\frac{t^5}{5}-580\frac{t^4}{4}+1044\frac{t^3}{3}-720\frac{t^2}{2}\right)f_1\right|_0^6$$

$$+15\left|\left(\frac{t^7}{7}-19\frac{t^6}{6}+137\frac{t^5}{5}-461\frac{t^4}{4}+702\frac{t^3}{3}-360\frac{t^2}{2}\right)f_2\right|_0^6$$

$$-20\left|\left(\frac{t^7}{7}-18\frac{t^6}{6}+121\frac{t^5}{5}-372\frac{t^4}{4}+508\frac{t^3}{3}-240\frac{t^2}{2}\right)f_3\right|_0^6$$

$$+15\left|\left(\frac{t^7}{7}-17\frac{t^6}{6}+107\frac{t^5}{5}-307\frac{t^4}{4}+396\frac{t^3}{3}-180\frac{t^2}{2}\right)f_4\right|_0^6$$

$$+6\left|\left(\frac{t^7}{7}-16\frac{t^6}{6}+95\frac{t^5}{5}-260\frac{t^4}{4}+324\frac{t^3}{3}-144\frac{t^2}{2}\right)f_5\right|_0^6$$

$$+\left|\left(\frac{t^7}{7}-15\frac{t^6}{6}+85\frac{t^5}{5}-225\frac{t^4}{4}+274\frac{t^3}{3}-120\frac{t^2}{2}\right)f_6\right|_0^6\Bigg)$$

$$=\frac{h}{720}[216f_0+1080f_1+216f_2+1296f_3+216f_4+1080f_5+216f_6]$$

$$=\frac{3h}{10}[f_0+5f_1+f_2+6f_3+f_4+5f_5+f_6]$$

By using Newton's forward difference formula:

According to Newton's forward difference formula (7.2.2), we have:

$$P_6(p) = f_0 + p\Delta f_0 + \frac{p(p-1)}{2!}\Delta^2 f_0 + \frac{p(p-1)(p-2)}{3!}\Delta^3 f_0 + \frac{p(p-1)(p-2)(p-3)}{4!}\Delta^4 f_0$$

$$+\frac{p(p-1)(p-2)(p-3)(p-4)}{5!}\Delta^5 f_0$$

$$+\frac{p(p-1)(p-2)(p-3)(p-4)(p-5)}{6!}\Delta^6 f_0$$

$$\int_{x_0}^{x_6} f(x)dx = \int_{x_0}^{x_6} P_6(x)dx$$

$$= h\int_0^6 \left(f_0 + p\Delta f_0 + \frac{p(p-1)}{2!}\Delta^2 f_0 + + \frac{p(p-1)(p-2)}{3!}\Delta^3 f_0 \right.$$

$$+\frac{p(p-1)(p-2)(p-3)}{4!}\Delta^4 f_0 + \frac{p(p-1)(p-2)(p-3)(p-4)}{5!}\Delta^5 f_0$$

$$\left. +\frac{p(p-1)(p-2)(p-3)(p-4)(p-5)}{6!}\Delta^6 f_0 \right) dp$$

$$= h\left(f_0 |p|_0^6 + \Delta f_0 \left|\frac{p^2}{2}\right|_0^6 + \frac{\Delta^2 f_0}{2}\left|\frac{p^3}{3} - \frac{p^2}{2}\right|_0^6 + \frac{\Delta^3 f_0}{6}\left|\frac{p^4}{4} - \frac{3p^3}{3} + \frac{2p^2}{2}\right|_0^6 + \frac{\Delta^4 f_0}{24}\left|\frac{p^5}{5} - 6\frac{p^4}{4} + 11\frac{p^3}{3} - 6\frac{p^2}{2}\right|_0^6 \right.$$

$$\left. + \frac{\Delta^5 f_0}{120}\left|\frac{p^6}{6} - 10\frac{p^5}{5} + 35\frac{p^4}{4} - 50\frac{p^3}{3} + 24\frac{p^2}{2}\right|_0^6 + \frac{\Delta^6 f_0}{720}\left|\frac{p^7}{7} - 15\frac{p^6}{6} + 85\frac{p^5}{5} - 225\frac{p^4}{4} + 274\frac{p^3}{3} - 120\frac{p^2}{2}\right|_0^6 \right)$$

$$= h\left(6f_0 + 18\Delta f_0 + 27\Delta^2 f_0 + 24\Delta^3 f_0 + \frac{123}{10}\Delta^4 f_0 + \frac{33}{10}\Delta^5 f_0 + \frac{42}{140}\Delta^6 f_0 \right)$$

$$= h\big(6f_0 + 18(f_1 - f_0) + 27(f_2 - 2f_1 + f_0) + 24(f_3 - 3f_2 + 3f_1 - f_0)$$

$$+\frac{123}{10}(f_4 - 4f_3 + 6f_2 - 4f_1 + f_0) + \frac{33}{10}(f_5 - 5f_4 + 10f_3 - 10f_2 + 5f_1 - f_0)$$

$$+\frac{41}{140}(f_6 - 6f_5 + 15f_4 - 20f_3 + 15f_2 - 6f_1 + f_0)\big)$$

$$= \frac{3h}{10}\ (20f_0 + 60\ (f_1 - f_0) + 90\ (f_2 - 2f_1 + f_0) + 80\ (f_3 - 3\ f_2 + 3f_1 - f_0) + 41\ (f_4 - 4f_3 + 6f_2 - 4f_1 + f_0) + 11(f_5 - 5f_4 + 10f_3 - 10f_2 + 5f_1 - f_0) + (f_6 - 6f_5 + 15f_4 - 20f_3 + 15f_2 - 6f_1 + f_0))$$

$$= \frac{3h}{10}\left[f_0 + 5f_1 + f_2 + 6f_3 + f_4 + 5f_5 + f_6\right] \qquad (7.9.2.1)$$

Equation (7.9.2.1) is Weddle's rule to find the approximate value of integrand f(x) numerically.

Example 7.10: Evaluate the integral $\int_1^e \frac{\sqrt{\log x}}{1+x}\, dx$ by using Weddle's rule.

Solution: According to Weddle's rule, we have:

$$\int_{x_0}^{x_6} f(x)dx = \frac{3h}{10}\left[f_0 + 5f_1 + f_2 + 6f_3 + f_4 + 5f_5 + f_6\right];$$

Now,

$$h = \frac{x_6 - x_0}{6} = \frac{e-1}{6} = 0.28638;$$

$x_0 = 1$,

$x_1 = 1 + 0.28638 = 1.28638$,

$x_2 = 1.28638 + 0.28638 = 1.57276$,

$x_3 = 1.57276 + 0.28638 = 1.85914$,

$x_4 = 1.85914 + 0.28638 = 2.14552$,

$x_5 = 2.14552 + 0.28638 = 2.43190$,

$x_6 = 2.43190 + 0.28638 = 2.71828$,

and $f(x) = \frac{\sqrt{\log x}}{1+x}$;

$$\Rightarrow f_0 = f(1) = \frac{\sqrt{\log 1}}{1+1} = 0,$$

$$\Rightarrow f_1 = f(1.28638) = \frac{\sqrt{\log 1.28638}}{1+1.28638} 0.21949,$$

$$\Rightarrow f_2 = f(1.57276) = \frac{\sqrt{\log 1.57276}}{1+1.57276} = 0.26156,$$

$$\Rightarrow f_2 = f(1.57276) = \frac{\sqrt{\log 1.57276}}{1+1.57276} = 0.26156,$$

$$\Rightarrow f_4 = f(2.14552) = \frac{\sqrt{\log 2.14552}}{1+2.14552} = 0.27777,$$

$$\Rightarrow f_5 = f(2.43190) = \frac{\sqrt{\log 2.43190}}{1+2.43190} = 0.27469;$$

$$\Rightarrow f_6 = f(2.71828) = \frac{\sqrt{\log 2.71828}}{1+2.71828} = 0.26894;$$

$$\therefore \int_{x_0}^{x_6} f(x)dx = \frac{3h}{10}[f_0 + 5f_1 + f_2 + 6f_3 + f_4 + 5f_5 + f_6]$$

$= = \frac{3}{10}$ 0 × 0.28638[1 × 0 + 5 × 0.21949 + 1 × 0.26156,6 × 0.27542 + 1 × 0.27777 + 5 × 0.27469 + 1 × 0.26894]

≈ 0.34972

Exercise

7.1. Find the roots of Legendre's polynomial of degree 2.

7.2. Find P_3 (x) and P_4 (x) where P_n (x) is Legendre's polynomial of degree n.

7.3. Find roots of P_3 (x) where P_n (x) is Legendre's polynomial of degree n.

7.4. Find Gauss-Legendre Translation for the following function:

$$\frac{1}{\pi}\int_0^{\pi} \cos(0.6 \text{ sint}) \text{ dt}$$

7.5. Find the approximate value of the following integral numerically by using the Gauss-Legendre two point Quadrature formula:

$$\int_{-1}^{1} \frac{\sin\left(\frac{a+1}{2}\right)}{a} \, da$$

7.6. Solve the following integral numerically by using the Gauss-Legendre four point Quadrature formula:

$$\int_{0}^{1} x^2 \, e^{x-1} \, dx$$

7.7. Euler-Meclaurin Formula

7.8. With the help of Trapezoidal Rule; find the approximate value of the integral:

$$4a\int_{0}^{\frac{\pi}{2}} \sqrt{1-\frac{(a+b)(a-b)\sin^2\theta}{a^2}} \, d\theta$$

7.9. Find the value of following integral by using Simpson's 1/3 Rule:

$$\int_{0}^{\frac{\pi}{2}} \frac{\cos x}{\sqrt{1+\sin x}} \, dx$$

7.10. Simplify the following integral with the help of Simpson's 3/8 Rule:

$$\int_{0}^{2} \frac{\sin t}{(t+1)^2 e^t} \, dt$$

7.11. Find the approximate value of the following series by using Euler-Maclaurin formula:

$$\frac{1}{51^2}+\frac{1}{53^2}+\frac{1}{55^2}+\ldots+\frac{1}{99^2}$$

7.12. Find the approximate value of the following integral by using Euler-Maclaurin formula:

$$\int_{0}^{1} \left(\frac{1}{x+1}\right) dx$$

7.13. Solve the following integral by using Boole's Rule:

$$\int_{0}^{2\pi} (5+2\sin x) \, dx$$

7.14. Find arc length of curve $y = f(x) = e^{-x^2}$ over $0 \le x \le 1$ by using Gauss-Legendre's two point formula. Length of arc is obtained by $L = \int_0^1 \sqrt{1+(f'(x))^2}\ dx$. Also, compare this value by approximating it with trapezoidal rule and Simpson's 1/3 rule.

7.15. The solid of revolution obtained by rotating the region under the curve y = f(x) = cos x, where $0 \le x \le \frac{\pi}{4}$ about the x-axis has surface area given by $2\pi \int_0^{\frac{\pi}{4}} f(x)\sqrt{1+(f'(x))^2}\ dx$ dx. Approx S by using Boole's rule.

7.16. Solve the following integral numerically by using Weddle's Rule:

$$\int_0^3 \sin^2\left(\frac{Z}{\pi}\right) dZ$$

8

Numerical Solution of Ordinary and Partial Differential Equations

8.1 Introduction

In the fields like Physics, Chemistry, Biology, Engineering and Economics; their laws and principles contains various parameters and their derivatives. The mathematical relationships of such parameters with their derivatives constitute the differential equations. In calculus, differential equations were firstly introduced by Newton and Leibnitz. Newton listed the three kinds of differential equations:

$$\frac{dy}{dx} = f(x), \quad \frac{dy}{dx} = f(x, y), \quad x_1 \frac{\partial y}{\partial x_1} + x_2 \frac{\partial y}{\partial x_2} = y$$

After that, Bernoulli, Euler, Lagrange, Fourier and many others introduced the various kinds of differential equations and their solution procedures. Most of differential equations are solved by using the analytic techniques. But in real life situations, these techniques do not give the satisfactory results. Thus, it is preferable for scientists and engineers to find the numerical approximation of their solutions in such circumstances. In this chapter, various numerical methods are explained to find the approximate solutions of differential equations. Some basic terms are described as follows:

Differential Equation

An equation of the form $\varphi(x, y, y', y'', ...) = 0$ is known as differential equation which contains independent variable x, dependendent variable y and its derivatives.

Ordinary Differential Equation

In a differential equation, if dependent variable depends on one independent variable only then corresponding differential equation is called as ordinary differential equation e.g.

a) $\frac{dT}{dt} = K(T_n - T)$ related to Newton's law of cooling

b) $m\frac{d^2y}{dt^2} + a\frac{dy}{dt} + K\,y = 0$ related to Simple Harmonic Motion

c) $L\frac{di}{dt} + iR = V$ related to Kirchhhoff's law

Order and degree of differential equation

The highest derivative present in differential equation is known as order of differential equation.

The power of term containing the highest order of derivative is known as degree of differential equation. But it should be non-negative and non-fractional. If it is not in such form then simplify the differential equation by treating as algebraically for converting the exponent of this term into required form.

Linear and Non-linear differential equations

A differential equation with degree one is known as linear differential equation if it does not contain the transcendental of dependent variable and also does not contain the product of dependent variable with its derivatives. If differential equation is not linear then it is called as non-linear differential equation.

Solution of differential equation

A solution of differential equation is a mathematical relation between dependent and independent variables which satisfies the given differential equation.

- **General Solution:** A solution of a differential equation which contains arbitrary constants is called as general solution. It contains the number of arbitrary constants equal to the order of differential equation e.g.

$$y = c_1e^{5x} + c_2e^{-5x} \text{ is general solution of } \frac{d^2y}{dx^2} - 25y = 0$$

- **Particular Solution:** A solution of a differential equation which contains no arbitrary constants is called as particular solution. It is derived from general solution of differential equation e.g.

$$y = e^{5x} + e^{-5x} \text{ is particular solution of } \frac{d^2y}{dx^2} - 25y = 0$$

One Step and Multistep Methods

In numerical methods for evaluating the approximate solution of differential equations, iterations are generally calculated which require the use of initial solution values. Those numerical methods which need the single initial value of solution are known as one-step methods for calculating the solution of differential equation. For example, Euler's method, Taylor's series method, Runge-Kutta method etc. While, the numerical methods which require the multiple initial values of solution are known as multi-step methods. For example, Adams-Bashforth method, Milne-Simpson's method etc.

Initial Value Problem

General solution of a differential equation contains the arbitrary constants. To find the value of such arbitrary constant, we need some initial conditions. When all such conditions are given at particular value of independent variable then the differential equation with such constraints is known as initial value problem. For example: $2y' + y - x = 0$, $y(0) = 1$; is an initial value problem.

Partial Differential Equation

In a differential equation, if dependent variable depends on more than one independent variable then corresponding differential equation is called as partial differential equation e.g.

$$\frac{\partial u}{\partial t} - K\left(\frac{\partial^2 u}{\partial x^2} + \frac{\partial^2 u}{\partial y^2} + \frac{\partial^2 u}{\partial z^2}\right) = 0; \quad = 0; \text{ (Laplace Equation)}$$

8.2 Picard's Method of Successive Approximation

Consider that we have a following differential equation whose solution is to be found:

$$\frac{dy}{dx} = f(x, y), \quad y(x_0) = y_0 \tag{8.2.1}$$

Integrate the equation (8.2.1) both sides to obtain the solution in the interval (x_0, x):

$$\Rightarrow \int_{x_0}^{x} dy = \int_{x_0}^{x} f(x, y)\, dx$$

$$\Rightarrow |y|_{x_0}^{x} = \int_{x_0}^{x} f(x, y)\, dx$$

$$\Rightarrow y(x) = y(x_0) + \int_{x_0}^{x} f(x, y)\, dx \tag{8.2.2}$$

Now, dependent variable y comes under the integral sign in right side of equation (8.2.2); therefore it is difficult to solve this equation. Thus, we have to replace y with some constant value and it can be taken as y_0 which gives the solution value as y_1. y_1 is called as the first approximate solution of equation (8.2.1) i.e.

$$y_1(x) = y_0 + \int_{x_0}^{x} f(x, y_0)\, dx$$

Further, y_1 can be used in equation (8.2.2) to find better approximation of solution and it is denoted by y_2 i.e.

$$y_2(x) = y_0 + \int_{x_0}^{x} f(x, y_1)\, dx$$

and this process continues till the required solution is obtained. Let y_i is an ith approximation and next approximated value of solution as y_{i+1} can be defined in the following way:

$$y_{i+1}(x) = y_0 + \int_{x_0}^{x} f(x, y_i)\, dx \tag{8.2.3}$$

Equation (8.2.3) is known as the Picard's Method of Successive Approximation for evaluating approximate solution of equation (8.2.1) numerically.

Example 8.1: Use Picard's Method of Succcssive Approximation to find the solution of following differential equation at $x = 0.2$:

$$\frac{dy}{dx} = xe^y; \quad y(0) = 0$$

Solution: According to Picard's Method of Successive Approximation, we have:

$$y_{i+1}(x) = y_0 + \int_{x_0}^{x} f(x, y_i)\, dx$$

Now, $x_0 = 0$, $y_0 = 0$, $f(x, y) = xe^y$

Computation of y_1:

$$\Rightarrow y_1(x) = y_0 + \int_{x_0}^{x} f(x, y_0)\, dx,$$

$$= 0 + \int_{0}^{x} x\, e^{0}\, dx$$

$$= \left| \frac{x^2}{2} \right|_0^x$$

$$= \frac{x^2}{2}$$

Computation of y_2:

$$\Rightarrow\ y_2(x) = y_0 + \int_{x_0}^{x} f(x, y_1)\, dx,$$

$$= 0 + \int_{0}^{x} x\, e^{\frac{x^2}{2}}\, dx$$

$$= e^{\frac{x^2}{2}} - 1$$

Now, it will be very difficult to solve next iteration. Therefore, we are stopping at y_2 :

$$\Rightarrow\ y_2(x) = e^{\frac{x^2}{2}} - 1$$

$$\Rightarrow y_2\,(0.2) = 0.02020$$

8.3 Taylor's Series Method

Recall equation (8.2.1) as a first order differential equation whose solution is to be find numerically. Let its solution is denoted by a curve y = F(x). Now we can expand y = F(x) about a point using $x = x_0$ Taylor's theorem of expansion. We have:

$$y(x) = y(x_0) + \frac{(x - x_0)}{1!} y'(x_0) + \frac{(x - x_0)^2}{2!} y''(x_0) + \frac{(x - x_0)^3}{3!} y'''(x_0) + \ldots + \frac{(x - x_0)^n}{n!} y^n(x_0) \tag{8.3.1}$$

where y', y" , …, y^n are first, second and n^{th} derivatives of y repectively.

Now, put $x_1 = x_0 + h$, $y = y_1$ and $y' = f$ in equation (8.3.1), we get:

$$y_1(h) = y_0 + \frac{(h)}{1!} f(x_0, y_0) + \frac{(h^2)}{2!} f'(x_0, y_0) + \frac{(h^3)}{3!} f''(x_0, y_0) + \ldots + \frac{(h^n)}{n!} f^{n-1}(x_0, y_0)$$

Now is the first approximation of solution. Further, it can be used in equation (8.2.2) to find next approximation i.e.

$$y_2(h) = y_1 + \frac{(h)}{1!} f(x_1, y_1) + \frac{(h^2)}{2!} f'(x_1, y_1) + \frac{(h^3)}{3!} f''(x_1, y_1) + \ldots + \frac{(h^n)}{n!} f^{n-1}(x_1, y_1)$$

and this process continues till the required solution is obtained. Let y_i is an ith approximation and next approximated value of solution as y_{i+1} can be defined in the following way:

$$y_{i+1}(h) = y_i + \frac{(h)}{1!} f(x_i, y_i) + \frac{(h^2)}{2!} f'(x_i, y_i) + \frac{(h^3)}{3!} f''(x_i, y_i) + \ldots + \frac{(h^n)}{n!} f^{n-1}(x_i, y_i) \quad (8.3.2)$$

Equation (8.3.2) is known as Taylor's Series Method for evaluating approximate solution of equation (8.2.1) numerically.

Example 8.2: Use Taylor's series expansion to obtain the approximate value of for the differential equation: $\frac{dy}{dx} = 3e^x + 2y, \quad y(0) = 0$.

Solution: According to Taylor's Method, we have:

$$y_{i+1} = y_i + \frac{(h)}{1!} f(x_i, y_i) + \frac{(h^2)}{2!} f'(x_i, y_i) + \frac{(h^3)}{3!} f''(x_i, y_i) + \ldots + \frac{(h^n)}{n!} f^{n-1}(x_i, y_i)$$

Now,

$f(x,y) = 3e^x + 2y$,

$\Rightarrow f'(x,y) = 3e^x + 2y' = 3e^x + 2f$,

$\Rightarrow f''(x,y) = 3e^x + 2f'$,

$\Rightarrow f'''(x,y) = 3e^x + 2f''$,

And $x_0 = 0$, $y_0 = 0$, take $h = 0.1$, $f(x, y) = 3e^x + 2y$

$\therefore x_1 = 0.1, \quad x_2 = 0.2$,

Computation of: y_1:

$\Rightarrow f(0,0) = 3, \quad f'(0,0) = 9, \quad f''(0,0) = 21, \quad f'''(0,0) = 45,$

$$\Rightarrow y_1 = 3\times\frac{(0.1)}{1!} + 9\times\frac{(0.1^2)}{2!} + 21\times\frac{(0.1^3)}{3!} + 45\times\frac{(0.1^4)}{4!} + \ldots,$$

$\Rightarrow y_1 = 0.3487$

Computation of y_2:

$\Rightarrow f(x_1, y_1) = 4.55, \quad f'(x_1, y_1) = 12.95, \quad f''(x_1, y_1) = 29.75, \quad f'''(x_1, y_1) = 63.36,$

$$\Rightarrow y_2 = 0.3487 + 4.55\times\frac{(0.1)}{1!} + 12.95\times\frac{(0.1^2)}{2!} + 29.75\times\frac{(0.1^3)}{3!} + 63.36\times\frac{(0.1^4)}{4!} + \ldots,$$

$\Rightarrow y_2 = 0.8736,$

Hence, y (0.2) = 0.8736

8.4 Euler's Method

Recall equation (8.2.1) as a first order differential equation whose solution is to be fond numerically. In this method, first two terms of Taylor's series expansion of equation (8.3.1) are taken i.e.

$$y(x) = y(x_0) + (x - x_0)\, y'(x_0) \tag{8.4.1}$$

and from equation (8.2.1), we have: $y'(x_0) = f(x_0, y_0)$;

use this value in equation (8.4.1); we get:

$$y(x) = y(x_0) + (x - x_0)\, f(x_0, y_0)$$

Put $x_1 = x_0 + h$ and y_1 is first approximate solution of equation (8.2.1) which is evaluated as following:

$$y_1 = y(x_0) + hf(x_0, y_0) \tag{8.4.2}$$

Now, it can be used further to find next approximation y_2 i.e.

$$y_2 = y(x_1) + hf(x_1, y_1)$$

and this process continues till the required solution is obtained. Let y_i is an ith approximation and next approximated value of solution as y_{i+1} can be defined in the following way:

$$y_{i+1} = y_i + hf(x_i, y_i) \tag{8.4.3}$$

Equation (8.4.3) is known as Euler's Method for evaluating approximate solution of equation (8.2.1) numerically.

Example 8.3: Given the equation: $\frac{dy}{dx} = 3x^2 + 2x, \quad y(1) = 2$ and estimate the value y(2) by taking $\frac{dy}{dx} = 3x^2 + 2x, \quad y(1) = 2$.

Solution: According to Euler's Method, we have:

$y_{i+1} = y_i + hf(x_i, y_i)$

Now, $x_0 = 1$, $y_0 = 2$, $h = 0.25$, $f(x, y) = 3x^2 + 2x$

$\therefore x_1 = 1.25, \quad x_2 = 1.5, \quad x_3 = 1.75, \quad x_4 = 2.0,$

Computation of : y_1:

$\Rightarrow y_1 = y_0 + hf(x_0, y_0),$

$\Rightarrow y_1 = 2 + 0.25(3(1)^2 + 2(1)),$

$\Rightarrow y_1 = 3.25,$

Computation of : y_2:

$\Rightarrow y_2 = y_1 + hf(x_1, y_1),$

$\Rightarrow y_2 = 3.25 + 0.25(3(1.25)^2 + 2(1.25)),$

$\Rightarrow y_2 = 5.0469,$

Computation of : y_3:

$\Rightarrow y_3 = y_2 + hf(x_2, y_2),$

$\Rightarrow y_3 = 5.0469 + 0.25(3(1.5)^2 + 2(1.5)),$

$\Rightarrow y_3 = 7.4844,$

Computation of : y_4:

$\Rightarrow y_4 = y_3 + hf(x_3, y_3),$

$\Rightarrow y_4 = 7.4844 + 0.25(3(1.75)^2 + 2(1.75)),$

$\Rightarrow y_4 = 10.6563,$

Hence, y(2)=10.6563

8.5 Modified Euler's Method

Euler method for evaluating the approximate solution of equation (8.2.1) has very low accuracy. It involves the large computational efforts and satisfactory solution is attained after the large number of iterations. It is modified on the basis of inserting average slope in equation (8.4.2) which is explained in the following way:

From Euler method, we recall the equation (8.4.2):

$$y_{11}(x) = y(x_0) + hf(x_0, y_0) \tag{8.5.1}$$

where y_1 is the first approximation of solution and next iteration is $x_1 = x_0 + h$. Now, we find the value of slope f(x,y) at the point (x_1, y_{11}) and the average value of slope at the points (x_0, y_0) and (x_1, y_{11}) is used in equation (8.5.1) for making correction in first approximate value y_{11}. Let the modified value of y_{11} is denoted by y_1 and it is evaluated as:

$$y_1(x) = y_0 + h\left(\frac{f(x_0, y_0) + f(x_1, y_{11})}{2}\right)$$

Now, we find the next approximate solution value y_2 by using Euler formula in the following way:

$$y_{21}(x) = y_1 + hf(x_1, y_1) \tag{8.5.2}$$

Similarly, next approximation is evaluated by using Euler's method and denoted this value as:

$$y_2(x) = y_1 + h\left(\frac{f(x_1, y_1) + f(x_2, y_{21})}{2}\right)$$

and this process continues till the required solution is obtained. Let y_{i1} is an ith approximation and next approximated value of solution as $y_{(i+1)1}$ is firstly calculated by using Euler's method in the following way:

$$y_{(i+1)1}(x) = y_i + hf(x_i, y_i) \tag{8.5.3}$$

and after that equation (8.5.3) is modified as:

$$y_{(i+1)} = y_i + h\left(\frac{f(x_i, y_i) + f\left(x_{i+1}, y_{(i+1)1}\right)}{2}\right) \tag{8.5.4}$$

Equation (8.5.4) is known as modified Euler's Method for evaluating approximate solution of equation (8.2.1) numerically.

Example 8.4: Given the equation: $\frac{dy}{dx} = \ln(x+y)$, $y(0)=1$ and estimate the value y(0.3) with h = 0.1 by using modified Euler's method.

Solution: According to modified Euler's Method, we have:

$$y_{(i+1)1}(x) = y_i + hf(x_i, y_i)$$

$$y_{(i+1)} = y_i + h\left(\frac{f(x_i, y_i) + f\left(x_{i+1}, y_{(i+1)1}\right)}{2}\right)$$

Now, $x_0 = 0$, $y_0 = 1$, $h = 0.1$, $f(x,y) = \log(x + y)$

$\therefore x_1 = 0.1, \quad x_2 = 0.2, \quad x_3 = 0.3,$

Computation of y_1:

$\Rightarrow y_{11} = y_0 + hf(x_0, y_0),$

$\Rightarrow y_{11} = 1,$

$$\Rightarrow y_1 = y_0 + h\left(\frac{f(x_0, y_0) + f(x_1, y_{11})}{2}\right),$$

$$\Rightarrow y_1 = 1 + 0.1\left(\frac{0+0}{2}\right),$$

$\Rightarrow y_1 = 1,$

Computation of y_2:

$\Rightarrow y_{21} = y_1 + hf(x_1, y_1),$

$\Rightarrow y_{21} = 1.010,$

$$\Rightarrow y_2 = y_1 + h\left(\frac{f(x_1, y_1) + f(x_2, y_{21})}{2}\right),$$

$$\Rightarrow y_2 = 1 + 0.1\left(\frac{0.953 + 0.104}{2}\right),$$

$\Rightarrow y_2 = 1.0099,$

Computation of y_3:

$\Rightarrow y_{31} = y_1 + hf(x_1, y_1)$,

$\Rightarrow y_{31} = 1.010$,

$$\Rightarrow y_3 = y_2 + h\left(\frac{f(x_2, y_2) + f(x_3, y_{31})}{2}\right),$$

$$\Rightarrow y_3 = 1.0099 + 0.1\left(\frac{0.1906 + 0.2062}{2}\right),$$

$\Rightarrow y_3 = 1.0298$,

Hence, $y(0.3) = 1.0298$,

8.6 Runge-Kutta Method

The German mathematicians Carl Runge and Martin Kutta developed a family of methods to find out the approximate solution of equation (8.2.1). These are one-step methods. Taylor's series method requires large computational efforts to find out the values of higher derivatives while in these methods there is no need to find higher derivatives. These methods are based on following General Euler formula:

$y_{i+1}(x) = y(x_i) + hf(x_i, y_i)$ or $y_{i+1}(x) = y(x_i) + h \times k$ (8.6.1)

where k is the value of slope of solution curve i.e.

New Iteration = Previous iteration + interval length × slope

These methods are named according to the number of points used for evaluating slope. Let the slope k is evaluated by using n points whose abscissa lie in the interval (x_i, x_{i+1}) and it can be defined as follows:

$K = c_1 k_1 + c_2 k_2 + ... + c_n k_n$

where $c_1, c_2, \ldots, c_n$ are the coefficients which give the weight-age to respective k_i's and k_i's are evaluated as follows:

$k_1 = f(x_i, y_i)$,

$k_2 = f(x_i + A_1 h, y_i + B_{11} k_1 h)$,

$k_3 = f(x_i + A_2 h, y_i + B_{21} k_1 h + B_{22} k_2 h)$,

....

$k_n = f(x_i + A_{n-1} h, y_i + B_{n-1}1 k_1 h + B_{n-1} 2 k_2 h + \ldots + B_{(n-1)(n-1)} k_{n-1} h)$,

Thus, equation (8.6.1) is written as:

$$y_{i+1}(x) = y(x_i) + h \times (c_1 k_1 + c_2 k_2 + \ldots\ldots. + c_n k_n) \quad (8.6.2)$$

Now, these c_i' s,A_i' s, B_{ij}' s (i = 1,2,...,n;j = 1,2,..., n-1) are computed by comparing equation (8.6.2) with the following Taylor's series expansion:

$$y(x+h) = y(x) + \frac{(h)}{1!} f(x,y) + \frac{(h)}{2!} f'(x,y) + \frac{(h)}{3!} f''(x,y) + \ldots + \frac{(h)}{n!} f^{n-1}(x,y)$$

(8.6.3)

These methods are named according to number of k_i's used in equation (8.6.2). Some of them are explained in the following way:

First Order Runge Kutta Method

In this method, only k_1is used and equation (8.6.2) is written as:

$$y_{i+1} = y(x_i) + h \times (c_1 k_1) \quad (8.6.4)$$

from equation (8.6.3), first order Taylor's series expansion is written as:

$$y(x_i + h) = y(x_i) + \frac{(h)}{1!} f(x_i, y_i) \quad (8.6.5)$$

Thus, after comparing (8.6.4) and (8.6.5); we get the value of $c_1 = 1$ and hence, equation (8.6.4) can be rewritten as:

$$y_{i+1} = y(x_i) + h \times (k_1) \quad (8.6.6)$$

where $k_1 = f(x_i, y_i)$

Equation (8.6.6) is known as first order Runge Kutta method to find a numerical approximate solution of equation (8.2.1).

Second Order Runge Kutta Method

In this method, k_1 and k_2 are used and equation (8.6.2) is written as:

$$y_{i+1} = y(x_i) + h \times (c_1 k_1 + c_2 k_2) \quad (8.6.7)$$

where, $k_1 = f(x_i, y_i)$ and $k_2 = f(x_i + A_1 h, y_i + B_{11} k_1 h)$,

from equation (8.6.3), second order Taylor's series expansion is written as:

$$y(x_i + h) = y(x_i) + \frac{(h)}{1!} f(x_i, y_i) + \frac{(h^2)}{2!} f'(x_i, y_i) \quad (8.6.8)$$

Thus, after comparing (8.6.7) and (8.6.8); we get the two sets of choice:

(a) $c_1 = \frac{1}{2},\ c_2 = \frac{1}{2},\ A_1 = 1,\ B_{11} = 1$

$$\therefore y_{i+1} = y(x_i) + h \times \left(\frac{k_1 + k_2}{2}\right) \quad (8.6.9)$$

where, $k_1 = f(x_i, y_i)$ and $k_2 = f(x_i + h, y_i + k_1 h)$,

Equation (8.6.9) is known as Heun's formula.

(b) $c_1 = \frac{1}{3},\ c_2 = \frac{2}{3},\ A_1 = \frac{1}{2},\ B_{11} = \frac{1}{2}$

$$\therefore y_{i+1} = y(x_i) + h \times \left(\frac{k_1 + 2k_2}{3}\right) \quad (8.6.10)$$

where, $k_1 = f(x_i, y_i)$ and $k_2 = f\left(x_i + \frac{h}{2},\ y_i + k_1 \frac{h}{2}\right)$

Equation (8.6.10) is also known as Modified Euler's formula.

Equations (8.6.9) and (8.6.10) are known as second order Runge Kutta method to find a numerical approximate solution of equation (8.2.1).

Third Order Runge Kutta Method

Similarly as explained above in two methods, after comparing the third order equations of (8.6.2) and (8.6.3); we get the following set of values:

$c_1 = \frac{1}{6},\ c_2 = \frac{4}{6},\ c_3 = \frac{1}{6},\ A_1 = \frac{1}{2},\ A_2 = 1, B_{11} = \frac{1}{2},\ B_{21} = -1,\ B_{22} = 2$

Thus, equation (8.6.2) can be written as:

$$\therefore y_{i+1} = y(x_i) + h \times \left(\frac{k_1 + 4k_2 + k_3}{6}\right) \quad (8.6.11)$$

where, $k_1 = f(x_i, y_i)$,

$k_2 = f\left(x_i + \frac{h}{2},\ y_i + k_1 \frac{h}{2}\right)$,

and $k_3 = f(x_i + h, y_i - k_1 h + 2k_2 h)$,

Equation (8.6.11) is known as third order Runge Kutta method to find a numerical approximate solution of equation (8.2.1).

Fourth Order Runge Kutta Method

Similarly as explained above in first two methods, after comparing the fourth order equations of (8.6.2) and (8.6.3); we get the following set of values:

$$c_1 = \frac{1}{6},\ c_2 = \frac{1}{3},\ c_3 = \frac{1}{3},\ c_4 = \frac{1}{6},\ A_1 = \frac{1}{2},\ A_2 = \frac{1}{2},\ A_3 = 1,$$

$$B_{11} = \frac{1}{2},\ B_{21} = 0,\ B_{22} = \frac{1}{2},\ B_{31} = 0,\ B_{32} = 0,\ B_{33} = 1,$$

Thus, equation (8.6.2) can be written as:

$$\therefore y_{i+1} = y(x_i) + h \times \left(\frac{k_1 + 2k_2 + 2k_3 + k_4}{6} \right) \tag{8.6.12}$$

where, $k_1 = f(x_i, y_i)$,

$$k_2 = f\left(x_i + \frac{h}{2},\quad y_i + k_1 \frac{h}{2} \right)$$

$$k_3 = f\left(x_i + \frac{h}{2},\quad y_i + k_2 \frac{h}{2} \right),$$

$k_4 = f(x_i + h, y_i + k_3 h)$,

Equation (8.6.12) is known as fourth order Runge Kutta method to find a numerical approximate solution of equation (8.2.1).

Example 8.5: Use first order Runge Kutta method to find the approximate solution of following differential equation at $x = 0.4$ with $h = 0.1$:

$$\frac{dy}{dx} = x^2 - y;\ y(0) = 1$$

Solution: According to first order Runge Kutta method, we have:

$y_{i+1} = y(x_i) + h \times (k_1)$ where $k_1 = f(x_i, y_i)$

Now, $x_0 = 0, h = 0.1, y_0 = 1$

therefore $x_1 = 0.1,\quad x_2 = 0.2,\quad x_3 = 0.3,\quad x_4 = 0.4,\quad f(x, y) = x^2 - y$

Computation of y_1:

$\Rightarrow k_1 = f(x_0, y_0) = x_0^2 - y_0 = -1,$

$\Rightarrow y_1 = y_0 + h \times (k_1) = 1 + 0.1 \times -1 = 0.9,$

Computation of y_2:

$\Rightarrow k_1 = f(x_1, y_1) = x_1^2 - y_1 = -0.89$

$\Rightarrow y_2 = y_1 + h \times (k_1) = 0.9 + 0.1 \times -0.89 = 0.811,$

Computation of y_3:

$\Rightarrow k_1 = f(x_2, y_2) = x_2^2 - y_2 = -0.771$

$\Rightarrow y_3 = y_2 + h \times (k_1) = 0.811 + 0.1 \times -0.771 = 0.7339,$

Computation of y_4:

$\Rightarrow k_1 = f(x_3, y_3) = x_3^2 - y_3 = -0.6439$

$\Rightarrow y_4 = y_3 + h \times (k_1) = 0.7339 + 0.1 \times -0.6439 = 0.66951,$

Hence, $y(0.4) = 0.66951$

Example 8.6: Use second order Runge Kutta method to find the approximate value of y(0.2), y(0.4), y(0.6) for the following differential equation:

$$\frac{dy}{dx} = y^2 + 1; \;\; y(0) = 0$$

Solution: According to second order Runge Kutta method, we have:

$$y_{i+1} = y(x_i) + h \times \left(\frac{k_1 + k_2}{2}\right)$$

where, $k_1 = f(x_i, y_i)$ and $k_2 = f(x_i + h, y_i + k_1 h)$

Now, $x_0 = 0$, $y_0 = 0$ and we take $h = 0.2$

therefore $x_1 = 0.2$, $x_2 = 0.4$, $x_3 = 0.6$, $f(x,y) = y^2 + 1$

Computation of y_1:

$\Rightarrow k_1 = f(x_0, y_0) = y_0^2 + 1 = 1,$

$\Rightarrow k_2 = f(x_0 + h, y_0 + k_1 h) = (y_0 + k_1 h)^2 + 1 = 1.04,$

$$\Rightarrow y_1 = y_0 + h \times \left(\frac{k_1 + k_2}{2}\right) = 0 + 0.2 \times \left(\frac{1 + 1.04}{2}\right) = 0.204,$$

Computation of y_2:

$\Rightarrow k_1 = f(x_1, y_1) = y_1^2 + 1 = 1.0416,$

$\Rightarrow k_2 = f(x_1 + h, y_1 + k_1 h) = (y_1 + k_1 h)^2 + 1 = 1.1700,$

$$\Rightarrow y_2 = y_1 + h \times \left(\frac{k_1 + k_2}{2}\right) = 0.204 + 0.2 \times \left(\frac{1.0416 + 1.1700}{2}\right) = 0.4252,$$

Computation of y_3:

$\Rightarrow k_1 = f(x_2, y_2) = y_2^2 + 1 = 1.1808,$

$\Rightarrow k_2 = f(x_2 + h, y_2 + k_1 h) = (y_2 + k_1 h)^2 + 1 = 1.4373,$

$$\Rightarrow y_3 = y_2 + h \times \left(\frac{k_1 + k_2}{2}\right) = 0.4252 + 0.2 \times \left(\frac{1.1808 + 1.4373}{2}\right) = 0.6870,$$

Hence, $y(0.2) = 0.204$, $y(0.4) = 0.4252$, $y(0.6) = 0.6870$,

Example 8.7: Use third order Runge Kutta method to find the approximate value of $y(1.4)$ for the following differential equation:

$$\frac{dy}{dx} = \frac{2y}{x};\ y(1) = 2$$

Solution: According to third order Runge Kutta method, we have:

$$y_{i+1} = y(x_i) + h \times \left(\frac{k_1 + 4k_2 + k_3}{6}\right) \tag{8.6.11}$$

where, $k_1 = f(x_i, y_i)$,

$$k_2 = f\left(x_i + \frac{h}{2},\quad y_i + k_1 \frac{h}{2}\right),$$

and $k_3 = f(x_i + h, y_i - k_1 h + 2k_2 h)$,

Now, $x_0 = 1$, $y_0 = 2$ and we take $h = 0.1$

therefore $x_1 = 1.1$, $x_2 = 1.2$, $x_3 = 1.3$, $x_4 = 1.4$, $f(x, y) = \frac{2y}{x}$

Computation of y_1:

$$\Rightarrow k_1 = f(x_0, y_0) = \frac{2y_0}{x_0} = 4,$$

$$\Rightarrow k_2 = f\left(x_0 + \frac{h}{2},\quad y_0 + k_1 \frac{h}{2}\right) = \frac{2\left(y_0 + k_1 \frac{h}{2}\right)}{x_0 + \frac{h}{2}} = 4.1905,$$

$$\Rightarrow k_3 = f(x_0 + h,\quad y_0 - k_1 h + 2k_2 h) = \frac{2(y_0 - k_1 h + 2k_2 h)}{x_0 + h} = 4.4329,$$

$$\Rightarrow y_1 = y_0 + h \times \left(\frac{k_1 + 4k_2 + k_3}{6} \right) = 2 + 0.1 \times \left(\frac{1 \times 4 + 4 \times 4.1905 + 1 \times 4.4329}{2} \right)$$

$= 2.4199,$

Computation of y_2:

$$\Rightarrow k_1 = f(x_1, y_1) = \frac{2y_1}{x_1} = 4.3998,$$

$$\Rightarrow k_2 = f\left(x_1 + \frac{h}{2}, \quad y_1 + k_1 \frac{h}{2} \right) = \frac{2\left(y_1 + k_1 \frac{h}{2} \right)}{x_1 + \frac{h}{2}} = 4.5911,$$

$$\Rightarrow k_3 = f(x_1 + h, \quad y_1 - k_1 h + 2k_2 h) = \frac{2(y_1 - k_1 h + 2k_2 h)}{x_1 + h} = 4.8303,$$

$$\Rightarrow y_2 = y_1 + h \times \left(\frac{k_1 + 4k_2 + k_3}{6} \right)$$

$$= 2.4199 + 0.1 \times \left(\frac{14.3998 \times +4 \times 4.5911 + 1 \times 4.8303}{2} \right) = 2.8798,$$

Computation of y_3:

$$\Rightarrow k_1 = f(x_2, y_2) = \frac{2y_2}{x_2} = 4.7997,$$

$$\Rightarrow k_2 = f\left(x_2 + \frac{h}{2}, \quad y_2 + k_1 \frac{h}{2} \right) = \frac{2\left(y_2 + k_1 \frac{h}{2} \right)}{x_2 + \frac{h}{2}} = 4.9917,$$

$$\Rightarrow k_3 = f(x_2 + h, \quad y_2 - k_1 h + 2k_2 h) = \frac{2(y_2 - k_1 h + 2k_2 h)}{x_2 + h} = 5.2280,$$

$$\Rightarrow y_3 = y_2 + h \times \left(\frac{k_1 + 4k_2 + k_3}{6} \right)$$

$$= 2.8798 + 0.1 \times \left(\frac{1 \times 4.7997 + 4 \times 4.9917 + 1 \times 5.2280}{2} \right) = 3.3797,$$

Computation of y_4:

$$\Rightarrow k_1 = f(x_3, y_3) = \frac{2y_2}{x_2} = 5.1996,$$

$$\Rightarrow k_2 = f\left(x_3 + \frac{h}{2},\quad y_3 + k_1\frac{h}{2}\right) = \frac{2\left(y_3 + k_1\frac{h}{2}\right)}{x_3 + \frac{h}{2}} = 5.3922,$$

$$\Rightarrow k_3 = f(x_3 + h,\quad y_3 - k_1 h + 2k_2 h) = \frac{2(y_3 - k_1 h + 2k_2 h)}{x_3 + h} = 5.6260,$$

$$\Rightarrow y_4 = y_3 + h \times \left(\frac{k_1 + 4k_2 + k_3}{6}\right)$$

$$= 3.3797 + 0.1 \times \left(\frac{1 \times 5.1996 + 4 \times 5.3922 + 1 \times 5.6260}{2}\right) = 3.9197,$$

Hence, $y(1.4) = 3.9197$,

Example 8.8: Use fourth order Runge Kutta method to find the approximate solution of following differential equation at $x = 0.4$ with $h = 0.1$:

$$\frac{dy}{dx} = x^2 + y^2; \quad y(0) = 1$$

Solution: According to fourth order Runge Kutta method, we have:

$$y_{i+1} = y_i + h \times \left(\frac{k_1 + 2k_2 + 2k_3 + k_4}{6}\right) \tag{8.6.12}$$

where, $k_1 = f(x_i, y_i)$,

$$k_2 = f\left(x_i + \frac{h}{2},\quad y_i + k_1\frac{h}{2}\right),$$

$$k_3 = f\left(x_i + \frac{h}{2},\quad y_i + k_2\frac{h}{2}\right),$$

$k_4 = f(x_i + h, y_i + k_3 h)$,

Now, $x_0 = 0$, $h = 0.1$, $y_0 = 1$

therefore $x_1 = 0.1$, $x_2 = 0.2$, $x_3 = 0.3$, $x_4 = 0.4$, $f(x,y) = x^2 + y^2$

Computation of y_1:

$$\Rightarrow k_1 = f(x_0, y_0) = x_0^2 + y_0^2 = 1,$$

$$\Rightarrow k_2 = f\left(x_0 + \frac{h}{2},\quad y_0 + k_1\frac{h}{2}\right) = \left(x_0 + \frac{h}{2}\right)^2 + \left(y_0 + k_1\frac{h}{2}\right)^2 = 1.105,$$

$$\Rightarrow k_3 = f\left(x_0 + \frac{h}{2},\quad y_0 + k_2\frac{h}{2}\right) = \left(x_0 + \frac{h}{2}\right)^2 + \left(y_0 + k_2\frac{h}{2}\right)^2 = 1.1161,$$

$$\Rightarrow k_4 = f(x_0 + h,\quad y_0 + k_3h) = (x_0 + h)^2 + (y_0 + k_3h)^2 = 1.2457,$$

$$\Rightarrow y_1 = y_0 + h\times\left(\frac{1\times1 + 2\times1.105 + 2\times1.1161 + 1\times1.2457}{6}\right) = 1.1115,$$

Computation of y_2:

$$\Rightarrow k_1 = f(x_1, y_1) = x_1^2 + y_1^2 = 1.2454,$$

$$\Rightarrow k_2 = f\left(x_1 + \frac{h}{2},\quad y_1 + k_1\frac{h}{2}\right) = \left(x_1 + \frac{h}{2}\right)^2 + \left(y_1 + k_1\frac{h}{2}\right)^2 = 1.4001,$$

$$\Rightarrow k_3 = f\left(x_1 + \frac{h}{2},\quad y_1 + k_2\frac{h}{2}\right) = \left(x_1 + \frac{h}{2}\right)^2 + \left(y_1 + k_2\frac{h}{2}\right)^2 = 1.4184,$$

$$\Rightarrow k_4 = f(x_1 + h,\quad y_1 + k_3h) = (x_1 + h)^2 + (y_1 + k_3h)^2 = 1.6108,$$

$$\Rightarrow y_2 = y_1 + h\times\left(\frac{1\times1.2454 + 2\times1.4001 + 2\times1.4184 + 1\times1.6108}{6}\right) = 1.2530,$$

Computation of y_3:

$$\Rightarrow k_1 = f(x_2, y_2) = x_2^2 + y_2^2 = 1.6100,$$

$$\Rightarrow k_2 = f\left(x_2 + \frac{h}{2},\quad y_2 + k_1\frac{h}{2}\right) = \left(x_2 + \frac{h}{2}\right)^2 + \left(y_2 + k_1\frac{h}{2}\right)^2 = 1.8408,$$

$$\Rightarrow k_3 = f\left(x_2 + \frac{h}{2},\quad y_2 + k_2\frac{h}{2}\right) = \left(x_2 + \frac{h}{2}\right)^2 + \left(y_2 + k_2\frac{h}{2}\right)^2 = 1.8717,$$

$$\Rightarrow k_4 = f(x_2 + h,\ y_2 + k_3\,h) = (x_2 + h)^2 + (y_2 + k_3\,h)^2 = 2.1641,$$

$$\Rightarrow y_3 = y_2 + h\times\left(\frac{1\times 1.6100 + 2\times 1.8408 + 2\times 1.8717 + 1\times 2.1641}{6}\right) = 1.4397,$$

Computation of y_4:

$\Rightarrow k_1 = f(x_3, y_3) = x_3^2 + y_3^2 = 2.1626,$

$$\Rightarrow k_2 = f\left(x_3 + \frac{h}{2},\quad y_3 + k_1\frac{h}{2}\right) = \left(x_3 + \frac{h}{2}\right)^2 + \left(y_3 + k_1\frac{h}{2}\right)^2 = 2.5182,$$

$$\Rightarrow k_3 = f\left(x_3 + \frac{h}{2},\quad y_3 + k_2\frac{h}{2}\right) = \left(x_3 + \frac{h}{2}\right)^2 + \left(y_3 + k_2\frac{h}{2}\right)^2 = 2.5735,$$

$$\Rightarrow k_4 = f\left(x_3 + h,\quad y_3 + k_3 h\right) = \left(x_3 + h\right)^2 + \left(y_3 + k_3 h\right)^2 = 3.0399,$$

$$\Rightarrow y_4 = y_3 + h\times\left(\frac{1\times 2.1626 + 2\times 2.5282 + 2\times 2.5735 + 1\times 3.0399}{6}\right) = 1.6961,$$

Hence, $y(0.4) = 1.6961$

8.7 Adams-Bashforth Method

It is a multistep method which requires the prior information of four solution points of equation (8.2.1). It is also a prediction-correction method which formulates two equations: one predicts the approximate value of solution and other corrects the approximate value of solution. This method can be derived from the fundamental theorem of calculus:

$$y_{i+1} = y\left(x_i\right) + \int_{x_i}^{x_{i+1}} f\left(x, y\right)dx \tag{8.7.1}$$

By using equation (8.7.1), we find both predictor and correction equations. First of all, we find the predictor equation for which we can use either Newton's or Lagrange's Polynomial approximation for f(x,y) based on: (x_{i-3}, f_{i-3}), (x_{i-2}, f_{i-2}), (x_{i-1}, f_{i-1}), (x_i, f_i) and integrate this function over $[x_i, x_{i+1}]$ in equation (8.7.1). Thus, we get the predictor solution which is denoted by p_{i+1} and defined as:

$$p_{i+1} = y_i + \frac{h}{24}\left[-9f_{i-3} + 37f_{i-2} - 59f_{i-1} + 55f_i\right] \tag{8.7.2}$$

Now, we find the corrector equation for which we can use either Newton's or Lagrange's Polynomial approximation for f(x, y) based on four solution points:

$$\left(x_{i-2}, f_{i-2}\right), \left(x_{i-1}, f_{i-1}\right), \left(x_i, f_i\right), \left(x_{i+1}, \overline{f}_{i+1}\right);$$

where $\overline{f}_{i+1} = f(x_{i+1}, p_{i+1})$ and integrate this function over $[x_i, x_{i+1}]$ in equation (8.7.1). Thus, we get the solution which is denoted by y_{i+1} and defined as:

$$y_{i+1} = y_i + \frac{h}{24}\left[f_{i-2} - 5f_{i-1} + 19f_i + 9\overline{f}_{i+1}\right] \tag{8.7.3}$$

Equation (8.7.3) is known as Adam-Bashforth method to find a numerical approximate solution of equation (8.2.1).

Example 8.9: Find y(0.6), if y(x) is the solution of differential equation $\frac{dy}{dx} = \frac{1}{2}(xy)$ by using Adams-Bashforth method and assume that

y(0) = 1,y (0.1) = 1.0025, y(0.2) = 1.0101, y(0.3) = 1.0228.

Solution: According to Adams-Bashforth method, we have:

$$p_{i+1} = y_i + \frac{h}{24}[-9f_{i-3} + 37f_{i-2} - 59f_{i-1} + 55f_i]$$

$$y_{i+1} = y_i + \frac{h}{24}\left[f_{i-2} - 5f_{i-1} + 19f_i + 9\overline{f}_{i+1}\right]$$

Now, we take h = 0.1,

and $x_{-3} = 0,\ x_{-2} = 0.1,\ x_{-1} = 0.2,\ x_0 = 0.3,\ x_1 = 0.4,\ x_2 = 0.5,\ x_3 = 0.6,$

$y_{-3} = 1,\ y_{-2} = 1.0025,\ y_{-1} = 1.0101,\ y_0 = 1.0228,$ and $f(x, y) = \frac{1}{2}(xy)$

Computation of y_1:

$$\Rightarrow f_{-3} = \frac{1}{2}(x_{-3} \times y_{-3}) = 0,$$

$$\Rightarrow f_{-2} = \frac{1}{2}(x_{-2} \times y_{-2}) = 0.0501,$$

$$\Rightarrow f_{-1} = \frac{1}{2}(x_{-1} \times y_{-1}) = 0.1010,$$

$$\Rightarrow f_0 = \frac{1}{2}(x_0 \times y_0) = 0.1534,$$

$$\Rightarrow p_1 = y_0 + \frac{h}{24}[-9f_{-3} + 37f_{-2} - 59f_{-1} + 55f_0] = 1.0409,$$

$$\Rightarrow \overline{f}_1 = \frac{1}{2}(x_1 \times p_1) = 0.2082,$$

$$\Rightarrow y_1 = y_0 + \frac{h}{24}\left[f_{-2} - 5f_{-1} + 19f_0 + 9\overline{f}_1\right] = 1.0409,$$

Computation of y_2:

$$\Rightarrow f_{-2} = \frac{1}{2}(x_{-2} \times y_{-2}) = 0.0501,$$

$$\Rightarrow f_{-1} = \frac{1}{2}(x_{-1} \times y_{-1}) = 0.1010,$$

$$\Rightarrow f_0 = \frac{1}{2}(x_0 \times y_0) = 0.1534,$$

$$\Rightarrow f_1 = \frac{1}{2}(x_1 \times y_1) = 0.2082,$$

$$\Rightarrow p_2 = y_1 + \frac{h}{24}\left[-9f_{-2} + 37f_{-1} - 59f_0 + 55f_1\right] = 1.0645,$$

$$\Rightarrow \overline{f}_2 = \frac{1}{2}(x_2 \times p_2) = 0.7823,$$

$$\Rightarrow y_2 = y_1 + \frac{h}{24}\left[f_{-1} - 5f_0 + 19f_1 + 9\overline{f}_2\right] = 1.0839,$$

Computation of y_3:

$$\Rightarrow f_{-1} = \frac{1}{2}(x_{-1} \times y_{-1}) = 0.1010,$$

$$\Rightarrow f_0 = \frac{1}{2}(x_0 \times y_0) = 0.1534,$$

$$\Rightarrow f_1 = \frac{1}{2}(x_1 \times y_1) = 0.2082,$$

$$\Rightarrow f_2 = \frac{1}{2}(x_2 \times y_2) = 0.7919,$$

$$\Rightarrow p_3 = y_2 + \frac{h}{24}\left[-9f_{-1} + 37f_0 - 59f_1 + 55f_2\right] = 1.2340,$$

$$\Rightarrow \overline{f}_3 = \frac{1}{2}(x_3 \times p_3) = 0.9170,$$

$$\Rightarrow y_3 = y_2 + \frac{h}{24}\left[f_0 - 5f_1 + 19f_2 + 9\overline{f}_3\right] = 1.1773,$$

Hence, y(0.6) = 1.1773

8.8 Prediction-Correction Method

In Euler modified method, we observed that accuracy in attaining the solution is improved by introducing a modified iteration in calculating any iteration of approximation. Recall the equations (8.5.3) and (8.5.4):

$$\text{P: } y_{(i+1)1} = y_i + hf(x_i, y_i) \tag{8.8.1}$$

$$\text{C: } y_{(i+1)} = y_i + h\left(\frac{f(x_i, y_i) + f\left(x_{i+1}, y_{(i+1)1}\right)}{2}\right) \tag{8.8.2}$$

Equation (8.8.1) predicts the approximate value of solution as $y_{(i+1)1}$ by using the previous approximation y_i. Therefore, this equation can be called as Predictor equation while equation (8.8.2) is used to improve the approximation $y_{(i+1)1}$ and this equation can be called as corrector equation. Thus, we can use both equations in evaluating single iteration and the methods including such equations are known as Predictor-Corrector methods. This method is also called as simple Predictor-corrector method which is constructed from Euler method (gives prediction) and Trapezoidal rule (corrects the prediction). Beside it, we are describing two more Predictor-Corrector methods. One is Adams-Bashforth method which is explained in section 8.7 and other is Milne's-Simpson's method which is described in the following way:

Milne's-Simpson's method: This method can be derived from the following equation:

$$y_{i+1} = y_i + \int_{x_{i-1}}^{x_{i+1}} f(x, y)\,dx \tag{8.8.3}$$

By using equation (8.8.3), we find both predictor and correction equations. First of all, we find the predictor equation for which we can use either Newton's or Lagrange's Polynomial approximation for f(x,y) based on (x_{i-3}, f_{i-3}), (x_{i-2}, f_{i-2}), (x_{i-1}, f_{i-1}), (x_i, f_i) and integrate this function over $[x_{i-3}, x_{i+1}]$ in equation (8.8.3). Thus, we get the predictor solution which is denoted by p_{i+1} and defined as:

$$p_{i+1} = y_{i-3} + \frac{4h}{3}\left[2f_{i-2} - f_{i-1} + 2f_i\right] \qquad (8.8.4)$$

Now, we find the corrector equation for which we can use either Newton's or Lagrange's Polynomial approximation for f(x, y) based on (x_{i-1}, f_{i-1}), (x_i, f_i), (x_{i+1}, f_{i+1});where $\overline{f}_{i+1}$ and integrate this function over $[x_i, x_{i+1}]$ in equation (8.8.3). Thus, we get the solution which is denoted by y_{i+1} and defined as:

$$y_{i+1} = y_i + \frac{h}{3}\left[f_{i-1} + 4f_i + \overline{f}_{i+1}\right] \qquad (8.8.5)$$

Equation (8.8.5) is known as Milne's-Simpson's method to find an approximate solution of equation (8.2.1).

Example 8.10: Find y(3), if y(x) is the solution of differential equation $\frac{dy}{dx} = \frac{1}{2}(x+y)$ by using Milne's-Simpson's method and assume that y(0) = 2, y(0.5) = 2.636, y(1) = 3.595, y(1.5) = 4.968.

Solution: According to Milne's-Simpson's method, we have:

$$p_{i+1} = y_{i-3} + \frac{4h}{3}\left[2f_{i-2} - f_{i-1} + 2f_i\right]$$

$$y_{i+1} = y_i + \frac{h}{3}\left[f_{i-1} + 4f_i + \overline{f}_{i+1}\right]$$

Now, we take h = 0.5,
and $x_{-3} = 0$, $x_{-2} = 0.5$, $x_{-1} = 1.0$, $x_0 = 1.5$, $x_1 = 2.0$, $x_2 = 2.5$, $x_3 = 3.0$,

$$y_{-3} = 2,\ y_{-2} = 2.636,\ y_{-1} = 3.595,\ y_0 = 4.968,\ \text{and } f(x, y) = \frac{1}{2}(x+y)$$

Computation of y_1:

$$\Rightarrow f_{-2} = \frac{1}{2}(x_{-2} + y_{-2}) = 1.568,$$

$$\Rightarrow f_{-1} = \frac{1}{2}(x_{-1} + y_{-1}) = 2.2975,$$

$$\Rightarrow f_0 = \frac{1}{2}(x_0 + y_0) = 3.234,$$

$$\Rightarrow p_1 = y_{-3} + \frac{4h}{3}[2f_{-2} - f_{-1} + 2f_0] = 6.871,$$

$$\Rightarrow \bar{f}_1 = \frac{1}{2}(x_1 + p_1) = 4.4355,$$

$$\Rightarrow y_1 = y_0 + \frac{h}{3}\left[f_{-1} + 4f_0 + \bar{f}_1\right] = 8.2462,$$

Computation of y_2:

$$\Rightarrow f_{-1} = \frac{1}{2}(x_{-1} + y_{-1}) = 1.568,$$

$$\Rightarrow f_0 = \frac{1}{2}(x_0 + y_0) = 2.2975,$$

$$\Rightarrow f_1 = \frac{1}{2}(x_1 + y_1) = 3.234,$$

$$\Rightarrow p_2 = y_{-2} + \frac{4h}{3}[2f_{-1} - f_0 + 2f_1] = 10.3741,$$

$$\Rightarrow \bar{f}_2 = \frac{1}{2}(x_2 + p_2) = 6.4371,$$

$$\Rightarrow y_2 = y_1 + \frac{h}{3}\left[f_0 + 4f_1 + \bar{f}_2\right] = 13.2734,$$

Computation of y_3:

$$\Rightarrow f_0 = \frac{1}{2}(x_0 + y_0) = 3.234,$$

$$\Rightarrow f_1 = \frac{1}{2}(x_1 + y_1) = 5.1231,$$

$$\Rightarrow f_2 = \frac{1}{2}(x_2 + y_2) = 7.8867,$$

$$\Rightarrow p_3 = y_{-1} + \frac{4h}{3}[2f_0 - f_1 + 2f_2] = 15.0072,$$

$$\Rightarrow \bar{f}_3 = \frac{1}{2}(x_3 + p_3) = 9.0036,$$

$$\Rightarrow y_3 = y_2 + \frac{h}{3}\left[f_1 + 4f_2 + \overline{f}_3\right] = 20.8857,$$

Hence, y (3) = 20.8857

8.9 Boundary Value Problems

When we solve the differential equations then its solution contains the arbitrary constants or integration constants and for calculating the value of solution, we need some constraints to evaluate such arbitrary constants. If these constraints describe the boundary conditions of a solution region then this type of differential equation with boundary conditions is known as boundary value problem. For example:

$$\frac{d^2y}{dx^2} + 25y = 0, \quad y(0) = 0, \quad y\left(\frac{\pi}{2}\right) = 1$$

8.10 Solution of Laplace Equation - Jacobi's Method, Gauss-Siedel Method

Laplace Equation:

An elliptic partial differential equation of the form:

$$\frac{\partial^2 u}{\partial x^2} + \frac{\partial^2 u}{\partial y^2} = \nabla^2 u = 0,$$

is known as Laplacc equation and the operator $\nabla^2 = \frac{\partial^2}{\partial x^2} + \frac{\partial^2}{\partial y^2}$ is called as Laplacian operator.

The value of u is often given at all the boundary points of rectangular region R in xy-Plane and its value at internal points of region is to be evaluated numerically. For this, Jacobi's and Gauss-Siedel methods are used to find its approximate values which are explained in the following way:

- **Jacobi's method**

Firstly, the region R is split into square grids of length h as shown in figure 8.1 and values of u at nodes where grid lines intersect are denoted by

$$u_{i,j} = u(x_i, y_j); for 1 \le i \le n,\ 1 \le j \le m .$$

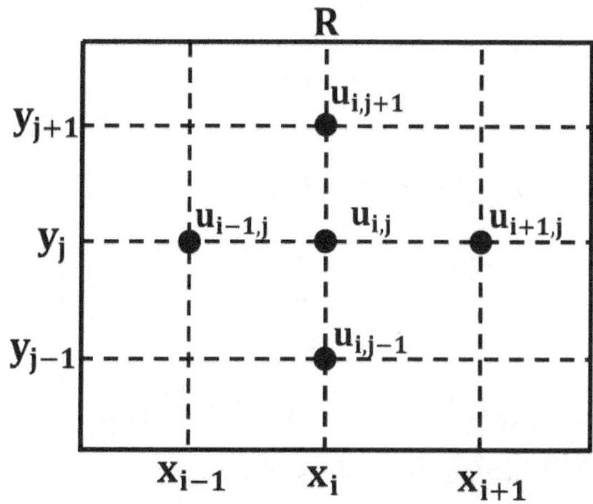

Fig. 8.1: Grid view of domain R

Consider that the boundary values are given at the following grid points:

$u(x_1, y_j) = u_{1,j} = \alpha$; *for* $2 \leq j \leq m-1$ (on the left side of region)

$u(x_i, y_1) = u_{i,1} = \beta$; *for* $2 \leq i \leq n-1$ (on the bottom of region)

$u(x_n, y_j) = u_{n,j} = \gamma$; *for* $2 \leq j \leq m-1$ (on the right side of region)

$u(x_i, y_m) = u_{i,m} = \delta$; *for* $2 \leq i \leq m-1$ (on the top of region)

where $\alpha, \beta, \gamma, \delta$ are numbers

Now, Laplace's difference equation for evaluating the value of u at ij-th node is applied. According to which, we have:

$$u_{i,j} = \frac{u_{i-1,j} + u_{i+1,j} + u_{i,j-1} + u_{i,j+1}}{4} \tag{8.10.1}$$

In this equation, values of u at five points are taken in which $\mu_{i,j}$ is taken at centre point which is non-boundary point and all remaining points are its up-down-left-right neighboring points as shown in the following figure 8.2:

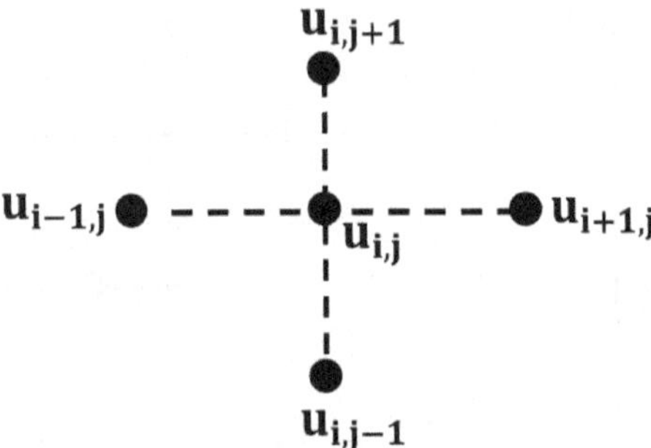

Fig. 8. 2: Grouping of neighboring points

In equation (8.10.1), we put the values of neighboring points accordingly as follows:

$$u_{i-1,j} = \begin{cases} \alpha, & if\ i = 2 \\ 0, & otherwise \end{cases},$$

$$u_{i,j-1} = \begin{cases} \beta, & if\ j = 2 \\ 0, & otherwise \end{cases},$$

$$u_{i+1,j} = \begin{cases} \gamma, & if\ i = n-1 \\ 0, & otherwise \end{cases},$$

$$u_{i,j+1} = \begin{cases} \delta, & if\ j = 2 \\ 0, & otherwise \end{cases},$$

Thus, we got first iteration of $u_{i,j}$ as $u_{i,j}^{1}$. In similar way, we got first iteration for all internal points of the domain. Now second iteration can be calculated as:

$$u^{(2)}_{i,j} = \frac{u_{i-1,j} + u_{i+1,j} + u_{i,j-1} + u_{i,j+1}}{4} \quad \text{where}$$

$$u_{i-1,j} = \begin{cases} \alpha, & if\ i = 2 \\ u^{(1)}_{i-1,j}, & otherwise \end{cases}$$

$$u_{i,j-1} = \begin{cases} \beta, & if\ j = 2 \\ u^{(1)}_{i,j-1}, & otherwise \end{cases}$$

$$u_{i+1,j} = \begin{cases} \gamma, & if\ i = n-1 \\ u^{(1)}_{i+1,j}, & otherwise \end{cases}$$

$$u_{i,j+1} = \begin{cases} \delta, & if\ j = 2 \\ u^{(1)}_{i,j+1}, & otherwise \end{cases}$$

and this process continues i.e. $u_{i,j}^{(i+1)}$ can be calculated as:

$$u^{(i+1)}_{i,j} = \frac{u_{i-1,j} + u_{i+1,j} + u_{i,j-1} + u_{i,j+1}}{4}$$

where

$$u_{i-1,j} = \begin{cases} \alpha, & if\ i = 2 \\ u^{(i)}_{i-1,j}, & otherwise \end{cases}$$

$$u_{i,j-1} = \begin{cases} \beta, & if\ j = 2 \\ u^{(i)}_{i,j-1}, & otherwise \end{cases}$$

$$u_{i+1,j} = \begin{cases} \gamma, & if\ i = n-1 \\ u^{(i)}_{i+1,j}, & otherwise \end{cases}$$

$$u_{i,j+1} = \begin{cases} \delta, & if\ j = 2 \\ u^{(i)}_{i,j+1}, & otherwise \end{cases}$$

This is the Jacobi's method to solve the Laplace equation numerically.

• Gauss-Siedel method

In this method, splitting of region R and grouping of grid points is done in similar way as explained in Jacobi's method. Also, Laplace's difference equation (8.10.1) for evaluating the value of u at ij-th node is used.

$$u_{i,j} = \frac{u_{i-1,j} + u_{i+1,j} + u_{i,j-1} + u_{i,j+1}}{4} \tag{8.10.2}$$

In equation (8.10.2), we put the values of neighboring points accordingly as follows:

$$u_{i-1,j} = \begin{cases} \alpha, & if\ i=2 \\ 0, & otherwise \end{cases},$$

$$u_{i,j-1} = \begin{cases} \beta, & if\ j=2 \\ 0, & otherwise \end{cases},$$

$$u_{i+1,j} = \begin{cases} \gamma, & if\ i=n-1 \\ 0, & otherwise \end{cases},$$

$$u_{i,j+1} = \begin{cases} \delta, & if\ j=2 \\ 0, & otherwise \end{cases},$$

Thus, we got first iteration of $u_{i,j}$ as $u_{i,j}^{1}$. Generally, domain is swept by using a pattern which is shown in figure 8.3:

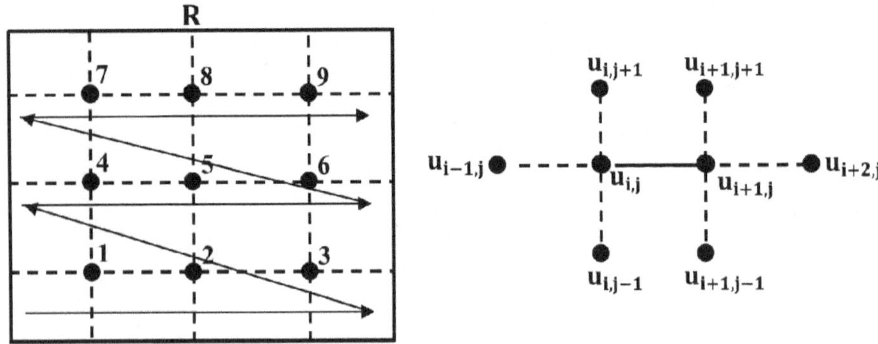

Fig. 8.3: Pattern for sweeping of domain R

Now, first iteration for u at next point $\left(u^{(1)}{}_{i+1,j}\right)$ of the domain is calculated as:

$$u^{(1)}{}_{i+1,j} = \frac{u^{(1)}{}_{i,j} + u_{i+2,j} + u_{i+1,j-1} + u_{i+1,j+1}}{4} \quad \text{where}$$

$$u_{i+1,j-1} = \begin{cases} \beta, & if\ j=2 \\ u_{i+1,j-1}, & otherwise \end{cases}$$

$$u_{i+2,j} = \begin{cases} \gamma, & if\ i=n-2 \\ u_{i+2,j}, & otherwise \end{cases}$$

$$u_{i+1,\, j+1} = \begin{cases} \delta, & \text{if } j = 2 \\ u_{i+1,\, j+1}, & \text{otherwise} \end{cases}$$

and this process continues i.e. $u_{i,j}^{(i+1)}$ can be calculated as:

$$u^{(i+1)}{}_{i,j} = \frac{u_{i-1,j} + u_{i+1,j} + u_{i,j-1} + u_{i,j+1}}{4}$$

where

$$u_{i-1,\, j} = \begin{cases} \alpha, & \text{if } i = 2 \\ u^{(i+1)}{}_{i-1,\, j}, & \text{otherwise} \end{cases}$$

$$u_{i,\, j-1} = \begin{cases} \beta, & \text{if } j = 2 \\ u^{(i)}{}_{i,\, j-1}, & \text{otherwise} \end{cases}$$

$$u_{i+1,\, j} = \begin{cases} \gamma, & \text{if } i = n-1 \\ u^{(i)}{}_{i+1,\, j}, & \text{otherwise} \end{cases}$$

$$u_{i,\, j+1} = \begin{cases} \delta, & \text{if } j = 2 \\ u^{(i)}{}_{i,\, j+1}, & \text{otherwise} \end{cases}$$

This is the Gauss-Siedel method to solve the Laplace equation numerically.

Example 8.11: Consider a steel plate of size 20cm × 20cm. If its left and lower sides are kept at 0 °C and other sides are kept at 100°C, Evaluate steady state temperature at interior points by using Jacobi's method and consider the grid size of 5cm × 5cm.

Solution: Now, According to Jacobi's method, we have:

$$u^{(i+1)}{}_{i,j} = \frac{u_{i-1,j} + u_{i+1,j} + u_{i,j-1} + u_{i,j+1}}{4}$$

where

$$u_{i-1,\, j} = \begin{cases} \alpha, & \text{if } i = 2 \\ u^{(i)}{}_{i-1,\, j}, & \text{otherwise} \end{cases}$$

$$u_{i,\, j-1} = \begin{cases} \beta, & if\ j = 2 \\ u^{(i)}{}_{i,\, j-1}, & otherwise \end{cases}$$

$$u_{i+1,\, j} = \begin{cases} \gamma, & if\ i = n-1 \\ u^{(i)}{}_{i+1,\, j}, & otherwise \end{cases}$$

$$u_{i,\, j+1} = \begin{cases} \delta, & if\ j = 2 \\ u^{(i)}{}_{i,\, j+1}, & otherwise \end{cases}$$

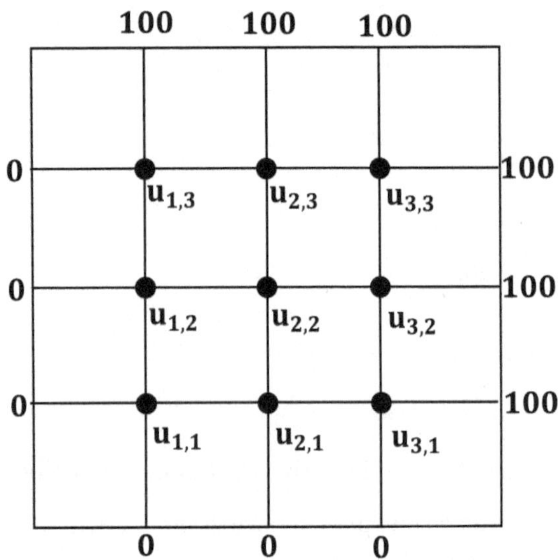

Therefore,

$$u_{1,1} = \frac{0 + u_{2,1} + 0 + u_{1,2}}{4},$$

$$u_{2,1} = \frac{u_{1,1} + u_{3,1} + 0 + u_{2,2}}{4},$$

$$u_{3,1} = \frac{u_{2,1} + 100 + 0 + u_{3,2}}{4},$$

$$u_{1,2} = \frac{0 + u_{2,2} + u_{1,1} + u_{1,3}}{4},$$

$$u_{2,2} = \frac{u_{1,2} + u_{3,2} + u_{2,1} + u_{2,3}}{4},$$

$$u_{3,2} = \frac{u_{2,2} + 100 + u_{3,1} + u_{3,3}}{4},$$

$$u_{1,3} = \frac{0 + u_{2,3} + u_{1,2} + 100}{4},$$

$$u_{2,3} = \frac{u_{1,3} + u_{3,3} + u_{2,2} + 100}{4},$$

$$u_{3,3} = \frac{u_{2,3} + 100 + u_{3,2} + 100}{4},$$

Various iterative values of temperature at internal grid points are listed in the table:

Example 8.12: Consider a steel plate of size 15cm×15cm. If its left and lower sides are kept at 0 °C and other sides are kept at 100 °C. Evaluate the steady state temperature at interior points by using Gauss-Siedel method. Assume that grid size is 5cm×5cm.

Solution: Now, According to Gauss-Siedel's method, we have:

$$u^{(i+1)}{}_{i,j} = \frac{u_{i-1,j} + u_{i+1,j} + u_{i,j-1} + u_{i,j+1}}{4}$$

where

$$u_{i-1,j} = \begin{cases} \alpha, & if\, i = 2 \\ u^{(i+1)}{}_{i-1,j}, & otherwise \end{cases}$$

$$u_{i,j-1} = \begin{cases} \beta, & if\, j = 2 \\ u^{(i)}{}_{i,j-1}, & otherwise \end{cases}$$

$$u_{i+1,j} = \begin{cases} \gamma, & if\, i = n-1 \\ u^{(i)}{}_{i+1,j}, & otherwise \end{cases}$$

$$u_{i,j+1} = \begin{cases} \delta, & if\, j = 2 \\ u^{(i)}{}_{i,j+1}, & otherwise \end{cases}$$

Table: 8.1: Value of temprature at various grid points

Iteration	$u_{1,1}$	$u_{2,1}$	$u_{3,1}$	$u_{1,2}$	$u_{2,2}$	$u_{3,2}$	$u_{1,3}$	$u_{2,3}$	$u_{3,3}$
1	0.0000	0.0000	0.0000	0.0000	0.0000	0.0000	0.0000	0.0000	0.0000
2	0.0000	0.0000	25.0000	0.0000	0.0000	25.0000	25.0000	25.0000	50.0000
3	0.0000	6.2500	31.2500	6.2500	12.5000	43.7500	31.2500	43.7500	62.5000
4	3.1250	10.9375	37.5000	10.9375	25.0000	51.5625	37.5000	51.5625	71.8750
5	5.4688	16.4063	40.6250	16.4063	31.2500	58.5938	40.6250	58.5938	75.7813
6	8.2031	19.3359	43.7500	19.3359	37.5000	61.9141	43.7500	61.9141	79.2969
7	9.6680	22.3633	45.3125	22.3633	40.6250	65.1367	45.3125	65.1367	80.9570
8	11.1816	23.9014	46.8750	23.9014	43.7500	66.7236	46.8750	66.7236	82.5684
9	11.9507	25.4517	47.6563	25.4517	45.3125	68.2983	47.6563	68.2983	83.3618
10	12.7258	26.2299	48.4375	26.2299	46.8750	69.0826	48.4375	69.0826	84.1492
11	13.1149	27.0096	48.8281	27.0096	47.6563	69.8654	48.8281	69.8654	84.5413
12	13.5048	27.3998	49.2188	27.3998	48.4375	70.2564	49.2188	70.2564	84.9327
13	13.6999	27.7903	49.4141	27.7903	48.8281	70.6472	49.4141	70.6472	85.1282
14	13.8951	27.9855	49.6094	27.9855	49.2188	70.8426	49.6094	70.8426	85.3236
15	13.9928	28.1808	49.7070	28.1808	49.4141	71.0379	49.7070	71.0379	85.4213
16	14.0904	28.2785	49.8047	28.2785	49.6094	71.1356	49.8047	71.1356	85.5190
17	14.1392	28.3761	49.8535	28.3761	49.7070	71.2333	49.8535	71.2333	85.5678
18	14.1881	28.4249	49.9023	28.4249	49.8047	71.2821	49.9023	71.2821	85.6166
19	14.2125	28.4738	49.9268	28.4738	49.8535	71.3309	49.9268	71.3309	85.6410
20	14.2369	28.4982	49.9512	28.4982	49.9023	71.3553	49.9512	71.3553	85.6655
21	14.2491	28.5226	49.9634	28.5226	49.9268	71.3797	49.9634	71.3797	85.6777
22	14.2613	28.5348	49.9756	28.5348	49.9512	71.3920	49.9756	71.3920	85.6899
23	14.2674	28.5470	49.9817	28.5470	49.9634	71.4042	49.9817	71.4042	85.6960

Contd.

24	14.2735	28.5531	49.9878	28.5531	49.9756	71.4103	49.9878	71.4103	85.7021
25	14.2766	28.5592	49.9908	28.5592	49.9817	71.4164	49.9908	71.4164	85.7051
26	14.2796	28.5623	49.9939	28.5623	49.9878	71.4194	49.9939	71.4194	85.7082
27	14.2811	28.5653	49.9954	28.5653	49.9908	71.4225	49.9954	71.4225	85.7097
28	14.2827	28.5669	49.9969	28.5669	49.9939	71.4240	49.9969	71.4240	85.7112
29	14.2834	28.5684	49.9977	28.5684	49.9954	71.4255	49.9977	71.4255	85.7120
30	14.2842	28.5691	49.9985	28.5691	49.9969	71.4263	49.9985	71.4263	85.7128
31	14.2846	28.5699	49.9989	28.5699	49.9977	71.4270	49.9989	71.4270	85.7131
32	14.2850	28.5703	49.9992	28.5703	49.9985	71.4274	49.9992	71.4274	85.7135
33	14.2851	28.5707	49.9994	28.5707	49.9989	71.4278	49.9994	71.4278	85.7137
34	14.2853	28.5709	49.9996	28.5709	49.9992	71.4280	49.9996	71.4280	85.7139
35	14.2854	28.5710	49.9997	28.5710	49.9994	71.4282	49.9997	71.4282	85.7140
36	14.2855	28.5711	49.9998	28.5711	49.9996	71.4283	49.9998	71.4283	85.7141

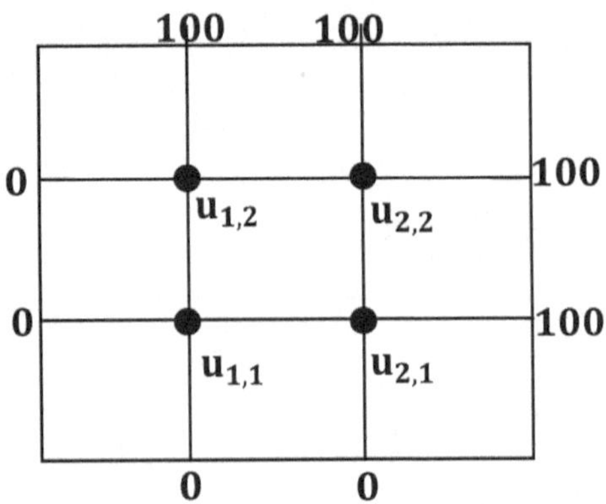

Therefore,

$$u_{1,1} = \frac{0 + u_{2,1} + 0 + u_{1,2}}{4},$$

$$u_{2,1} = \frac{u_{1,1} + 100 + 0 + u_{2,2}}{4},$$

$$u_{1,2} = \frac{0 + u_{2,2} + u_{1,1} + 100}{4},$$

$$u_{2,2} = \frac{u_{1,2} + 100 + u_{2,1} + 100}{4},$$

Now, first iterations are:

$$u^{(1)}{}_{1,1} = \frac{0 + u^{(0)}{}_{2,1} + 0 + u^{(0)}{}_{1,2}}{4} = \frac{0+0+0+0}{4} = 0,,$$

$$u^{(1)}{}_{2,1} = \frac{u^{(1)}{}_{1,1} + 100 + 0 + u^{(0)}{}_{2,2}}{4} = \frac{0+100+0+0}{4} = 25,$$

$$u^{(1)}{}_{1,2} = \frac{0 + u^{(0)}{}_{2,2} + u^{(1)}{}_{1,1} + 100}{4} = \frac{0+0+0+100}{4} = 25,$$

$$u^{(1)}{}_{2,2} = \frac{u^{(1)}{}_{1,2} + 100 + u^{(1)}{}_{2,1} + 100}{4} = \frac{25+100+25+100}{4} = 62.5,$$

Continuing in this way, various iterative values of temperature at internal grid points are listed in the table 8.2:

Table 8.2: Value of temprature various internal grid points

Iteration	$u_{1,1}$	$u_{2,1}$	$u_{1,2}$	$u_{2,2}$
1	0.0000	25.0000	25.0000	62.5000
2	12.5000	43.7500	43.7500	71.8750
3	21.8750	48.4375	48.4375	74.2188
4	24.2188	49.6094	49.6094	74.8047
5	24.8047	49.9023	49.9023	74.9512
6	24.9512	49.9756	49.9756	74.9878
7	24.9878	49.9939	49.9939	74.9969
8	24.9969	49.9985	49.9985	74.9992
9	24.9992	49.9996	49.9996	74.9998
10	24.9998	49.9999	49.9999	75.0000
11	25.0000	50.0000	50.0000	75.0000
12	25.0000	50.0000	50.0000	75.0000

Hence,

$u_{1,1} = 25, u_{2,1} = 50, u_{1,2} = 50, u_{2,2} = 50$

Exercise

8.1. Use Taylor's series expansion to solve the following differential equations:

(a) $\frac{dy}{dx} = x + y + xy, \ \ y(0) = 1;$ for x=0.4

(b) $\frac{dy}{dx} = x^2 y^2, \ \ y(1) = 0;$ for x=2.5

8.2. Use Euler's method to solve the following differential equations

(a) $\frac{dy}{dx} = x + y + xy, \ \ y(0) = 1;$ for x=0.4

(b) $\frac{dy}{dx} = x^2 y^2, \ \ y(1) = 0;$ for x=2.5

8.3. Solve the differential equations of Q. No. 8.2 by using modified Euler's method and compare the results.

8.4. Find the solution of differential equation: $\frac{dy}{dx} = y + \sin x, \ \ y(0) = 2;$ by using first order Runge-Kutta method.

8.5. Find the solution of differential equation: $\frac{dy}{dx} = y + \cos x, \ \ y(0) = 1;$ by using second order Runge-Kutta method.

8.6. Solve: $\frac{dy}{dx} = y + \sqrt{y},\ y(0) = 1$; by using simple Predictor-Corrector method.

8.7. Use Adams-Bashforth and Milne's-Simpson's method to compute approximate solution of differential equation: $\frac{dy}{dx} = y + \sqrt{y}, \quad y(0) = 1$; over [0,5] such that y(0.05) = 1.0550422, y(0.10) = 1.1203418, y(0.15) = 1.1961685.

8.8. Find the solution of differential equation: $\frac{dy}{dx} = e^{-2x} - e^{y}, \quad y(0) = \frac{1}{10}$; by using fourth order Runge-Kutta method.

8.9. Find the solution of differential equation: $\frac{dy}{dx} = 2xy^2, \quad y(0) = 1$; by using third order Runge-Kutta method.

8.10. Find the solution of differential equation: $\frac{dy}{dx} = \frac{2xy + e^x}{x^2 + xe^x}, \quad y(1) = 0$; by using second order Runge-Kutta method.

8.11. Use 5 × 5 grid to determine the approximate values of harmonic function u (x, y) by using Jacobi's method in rectangle R={*(x, y): $0 \le x \le 1, 0 \le y \le 1$*}. The boundary values are:

$$u(x,0) = x^2,\ u(x,1) = (x-1)^2, \quad for\ 0 \le x \le 1,$$

$$u(0,y) = y^2,\ u(1,y) = (y-1)^2, \quad for\ 0 \le y \le 1,$$

8.12. Use 6 × 6 grid to determine the approximate values of harmonic function u(x,y) by using Gauss-Siedel's method in rectangle R={(x,y): *$0 \le x \le 6$, $0 \le y \le 6$*}. The boundary values are:

$$u(x,0) = 0,\ u(x, 6) = 120, \text{ for } 0 \le x \le 6,$$

$$u(0,y) = 0, u(1,y) = (y - 1)^2, \text{ for } 0 \le y \le 6,$$

Index

A

Absolute Errors 13
Accuracy 13
Adams-Bashforth Method 270
Adjoint of a Square Matrix 31
Aitken Exploration 123
Aitken's D2 Method 123
Algebraic equation 93
Analytical methods 1
Argument 147
Augmented Matrix 42
Average operator 154

B

Backward Differences 149
backward differences operator 149
band matrix 66
Bessel's Formula 176
Binary Chopping 96
Binary Number System 6
Bisection (or Bolzano) Method 95
Boole's Rule 230
Boundary Value Problems 276
By using Lagrange's Interpolation Formula 215

C

Central Difference Interpolation Formula 172
Central difference operator 151
Central Differences 151
Chance errors 16
Chebyshev Method 122
Cholesky's Triangularisation Method 62
Clamped boundary condition 181
Cofactor 26
Column matrix 28
Column vector 28
Comparison of Iterative Methods 125
Complete pivoting 46
Convergence of Newton-Raphson Method 107
Convergence of the Secant Method 118
Convergence property of Newton's method 108
Conversion 7, 8
Cramer's Rule 35
Crout's Method 55
Cubic Spline Interpolation 178
Cubic spline interpolation 179
curve fitting 187
Curve Fitting with Linear 191
Curve Fitting with Linear Equation 191

D

Decimal Number System 6
Derivatives Based on Bessel's Interpolation Formul 141
Derivatives Based on Newton's Backward Interpolati 135
Derivatives Based on Newton's Divided Difference F 143
Derivatives Based on Newton's Forward Interpolatio 133
Derivatives Based on Stirling's Interpolation Form 138
Descarte's Rule of Signs 94
Determinants 25
Diagonal matrix 29
Differential Equation 251
Differential operator 154
Direct methods 32
Doolittle's method 55

E

Elementary matrix 32
Elementary row operations 42
Elementary transformations of a matrix 31
Entry 147
Equivalent matrix 32
Error 13
Error Estimate in Simpson's 1/3 Rule 222
Error Estimate in Trapezoidal Rule 217
Error Estimation 18
Error Propagation 16
Error Propagation in Difference Table 152
Euler-Meclaurin Formula 212
Euler's Method 257
Extended form of Newton-Raphson formula 123
Extrapolation 161

F

Factorial Notation 156
Factorization method 55
Finite Difference Operators 147
First Order Runge Kutta Method 262
Fitting of exponential curve 197
Fitting of straight line 194
Fixed point method 104
Forward difference operator 148
Forward Differences 147
Fourth Order Runge Kutta Method 264
Function Approximation 21

G

Gauss Elimination Method 44
Gauss Forward Interpolation Formula 172
Gauss-Jordan Elimination 43
Gauss-Jordan Method 52
Gauss-Legendre one point rule 205
Gauss-Legendre Quadrature Formulae 205
Gauss-Legendre two point rule 206
Gauss-Seidal Iteration Method 77
Gauss-Siedel method 279
Gaussian Integration 211
General Error Formula 19
Graphical methods 1

H

Heun's formula 263
Hexadecimal Number System 7
Human error 1

I

Incremental Search Method 96
Inherent Errors 14
Initial Value Problem 253
Intemediate Value Theorem 93
Interpolating function 161
Interpolating polynomial 161
Interpolation 161
Interpolation with Equal Intervals 161
Interpolation with Unequal Intervals 167
Interval Halving method 96
Inverse of a Matrix 31
Iteration Method 103, 126
Iterative Method for Eigen Values and Eigen Vector 82
Iterative methods 32

J

Jacobi's Iteration Method 71
Jacobi's method 276
Jacobi's Method 83

K

Kepler's Law 187
Knots 179

L

Lagrange's Interpolation Formula 167
Laplace Equation 276
Laplace-Everett's Formula 178
Laplace's difference equation 277
leading diagonal 25, 28
Legendre's polynomial 208
Linear and Non-linear differential equations 252
Linear Equation 188
Lower bandwidth 67
Lower Traingular matrix 29
LU Decomposition method 55

M

Matrices 27
Matrix Inversion Method 38
Method of Least Squares Fitting 194
Method of simultaneous displacements 71
Method of successive displacements 77
Method of Variable Secant 99
Milne's-Simpson's method 273, 274
Milne's-Simpson's method: 273
Modified Euler's Method 259
Modified Newton-Raphson Method 111
Muller's Method 119

N

Natural (or free) boundary condition 181
Natural cubic 179
Newton-Cotes Closed Quadrature Formula 203
Newton-Gregory formula for backward interpolation 166
Newton-Gregory formula for forward interpolation w 163
Newton-Raphson Method 104
Newton's backward difference interpolation formula 166
Newton's Backward Interpolation Formula 165
Newton's forward difference interpolation formula 163
Newton's Forward Interpolation Formula 161
Newton's General Interpolation Formula with Divide 169
Non-Linear Equations 126
Normal form of a matrix 32
Null matrix 29
Number System 5
Numerical methods 1

O

Octal Number System 7
Ohm's law 187
One Step and Multistep Methods 253
Operations on matrices 30
Order and degree of differential equation 252
Order of Convergence of Newton Raphson Method 109
Ordinary Differential Equation 251
Other Difference Operator 154

P

Partial Differential Equation 253
Partial pivoting 46
Picard's Method of Successive Approximation 253
Piecewise polynomial approximation 178
Pivoting 45
Polynomial interpolation 161
Prediction-Correction Method 273
Properties of determinants 26

Q

Quadratic convergence 115
Quadrature formula 202

R

Random Errors 16
Rank of a Matrix 31
Rate of Convergence of Modified Newton-Raphson Met 113
Rate of Convergence of Newton Raphson Method 110
Recursive Algorithms for Conversion 11
Regression analysis 187
Regula Falsi or False Position Method 98
Relationship between the Operators 154
Relative Error 202
Relative Errors 14
Reyleigh's Power Method 87
Round-off Errors 14
Row matrix 27
Row Vector 28
Rules for counting significant figures 12
Rules of rounding off 12
Runge-Kutta Method 261

S

Secant Method 116
Second Order Runge Kutta Method 262
Shift operator 154
Shortcut Method for Conversion 10
Significant Digits 11
Simpson's 1/3 Rule 218

Simpson's 3/8 Rule 225
Simultaneous equations 25
Skew-symmetric matrix 29
Smoothing function 161
Solution of differential equation 252
Spline 179
Spline interpolation 179
Square matrix 28
Square root method 62
Stability and Condition 22
Stable 23
Stirling's Formula 176
Sub-diagonal 67
Submatrix 28
Successive Approximation Method 114
Super-diagonal 67
Symmetric matrix 67
Systematic Errors 16

T

Taylor's Series Method 255
Taylor's Theorem 4
The Inverse of a Matrix 33
Third Order Runge Kutta Method 263
Thomas' algorithm 67
Thomas Algorithm for Tridiagonal System 66
Total error 18
Trace of the matrix 28
Transcendental equation 93
Transcendental Equations 94
Transpose of Matrix 30
Trapezoidal Rule 214
Tridiagonal matrices 30
Tridiagonal matrix 67
Tridiagonal matrix algorithm (TDMA) 67
Tridiagonal system of equations 67
Trigonometric interpolation 161
Truncation Errors 15

U

Uncertainty in Data or Noise 23
Unit matrix 29
Unstable 23
Upper bandwidth 67
Upper Traingular matrix 29

V

Vector 28

W

Weddle's Rule 236

Z

Zero matrix 30

Zeitfracht Medien GmbH
Ferdinand-Jühlke-Straße 7
99095 Erfurt, Deutschland
produktsicherheit@kolibri360.de